THE BEHAVIOUR (

Biological Principles and Implications for Production

J.J. Lynch

CSIRO Division of Animal Production
Private Mail Bag
Armidale NSW 2350
Australia

G.N. Hinch

Department of Animal Science
University of New England
Armidale NSW 2351
Australia

D.B. Adams

Bureau of Rural Resources
P.O. Box 858
Canberra 2601
Australia

C·A·B International

and

CSIRO
AUSTRALIA

5724   2320

10/97

LLYFRGELL
LLYSFASI
12064 LIBRARY 636.3

C·A·B International   Tel: Wallingford (0491) 32111
Wallingford   Telex: 847964 (COMAGG G)
Oxon OX10 8DE   Telecom Gold/Dialcom: 84: CAU001
UK   Fax: (0491) 33508

Published in Australia and New Zealand by:

CSIRO Publications
314 Albert St
East Melbourne
Victoria 3002
Australia

Tel: (03) 418 7217   Int + (613) 418 7217
Fax: (03) 419 0459   Int + (613) 419 0459

© C·A·B International 1992. All rights reserved. No part of this
publication may be reproduced in any form or by any means,
electronically, mechanically, by photocopying, recording or
otherwise, without the prior permission of the copyright owners.

A catalogue record for this book is available from the British Library

CABI   ISBN 0 85198 787 7
CSIRO ISBN 0 643 05329 8

Typeset by Leaper & Gard Ltd, Bristol, England
Printed and bound in the UK by Redwood Press Ltd, Melksham

Contents

Preface

The justification for this book lies with the twin aims of understanding behaviour by examining its biological basis and improving the husbandry of sheep by realizing that prior experience and the sheep's conservative yet undemonstrative behaviour must be considered in all management decisions. Sheep show these behaviours the first time they experience yards, a shed, new paddocks or even a new food. The following cartoon shows the problems of the unthinking manager. Sheep are not so much slow as wary of different situations.

The chapters which follow present the biological basis for each of the main behavioural areas in a sheep population. Thus Chapter 1 considers the issues relating to the maintenance of life through acquisition of food. Chapter 2 discusses the social organization of sheep which to some degree controls access to food resources. This social organization is involved in the maintenance of contact between the sexes, the topic of Chapter 3. Reproductive behaviour leads to pregnancy, to the expression of maternal care

and to the survival and development of the lamb which eventually takes its place with the adult flock. These topics are treated in Chapters 4 and 5. Finally, Chapter 6 provides a theoretical framework for welfare concerns and evaluates how our knowledge of behaviour may be used to assist mankind in maintaining the well-being of sheep under their control.

We would like to acknowledge with gratitude the following people who have read several versions of these chapters: Drs Manika Cockrem, Peter Jarman, John Nolan, Fred Provenza, John Thwaites, as well as Geoff Green, Judy Hinch, Margaret and Cathie Lynch. They have all assisted us materially with their comments, advice and encouragement, even when reading early drafts of this book. We thank Geoff Green for drawing and printing figures. Finally, we would like to thank Faye Hughes for her general assistance, advice, tolerance and skills in typing the text.

Several scientists have had a large influence on our thinking about ethology over the years. We acknowledge Ron Kilgour, David Lindsay, Glen McBride, Alex Stolba, Margaret Vince, Elizabeth Walser and David Wood-Gush for their contribution to the understanding of sheep behaviour. The emphasis placed on studying a species in relation to soils, plants and other components of the environment by H.G. Andrewartha, G.L. McClymont and W.M. Willoughby is consistent with the study of ethology and has moulded our approach to agricultural production.

Introduction

The genus *Ovis* is widely distributed throughout the world and is particularly concentrated in the temperate latitudes of the northern and southern hemispheres. Figure I.1(a) illustrates this distribution and also indicates the ten countries with the largest sheep populations. Archaeological evidence has shown that sheep were one of the first animal species domesticated by mankind, probably some 11,000 to 12,000 years ago. Since that time man has selected sheep for a wide variety of anatomical and production characteristics. There are now over 2000 breeds throughout the world.

The genus has species which vary in diploid number of chromosomes, in coat type and colour, in the size, number and presence of horns, in tail shape and body size (40 to 200 kg). However, the classification of what constitutes a species is far from clear and varies from author to author. It appears that all species will interbreed in captivity. To a degree, species classifications are based on differences in social behaviours and regional location rather than anatomical differences.

The wild species identified include the Bighorn (*Ovis canadensis*) of North America, the Argali (*Ovis ammon*) of central Asia, the Urial (*Ovis vignei*) of south-western Asia and the Mouflon (*Ovis musimon*) of western Asia and Europe. A number of other species have also been identified. However, those mentioned above are the most important in terms both of existing numbers and as the source of animals from which domestic sheep came (Ryder, 1984).

The wild species are all found in mountainous or high plains country. It has been suggested that the origins of the domestic sheep were in western Asia particularly in the high plains regions of what is now Turkey, Iran and Iraq. It is thought that the Urial was probably the initial wild species involved in domestication. However, the Argali is likely to have contri-

1

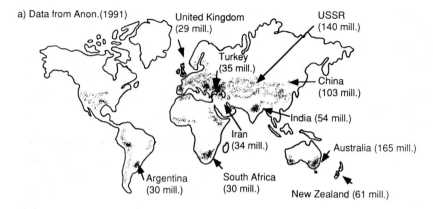

a) Data from Anon.(1991)

United Kingdom
(29 mill.)

USSR
(140 mill.)

Turkey
(35 mill.)

China
(103 mill.)

India (54 mill.)

Iran
(34 mill.)

Australia (165 mill.)

Argentina
(30 mill.)

South Africa
(30 mill.)

New Zealand (61 mill.)

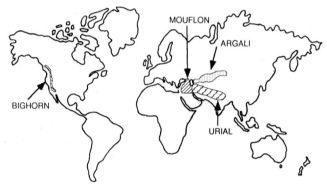

b) Data from Isaac (1970)

MOUFLON

ARGALI

BIGHORN

URIAL

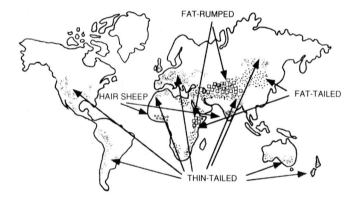

c) Data from Mason (1969)

FAT-RUMPED

HAIR SHEEP

FAT-TAILED

THIN-TAILED

Fig. I.1. The distribution of sheep throughout the world for (a) total numbers; (b) wild species; and (c) tail types.

buted to the development of the present domestic Asiatic breeds and the Mouflon to the European breeds. Figure I.1(b) shows the present world distribution of the four main wild sheep species and illustrates the high concentration of three of the wild species in the region where evidence for the original domestication of sheep has been found. A more detailed account of the origins of the domestic sheep and also of the more recent history of sheep and their relationship to mankind can be found in Ryder (1983).

The breeds of domesticated sheep which now constitute the species *Ovis aries* are more diverse in anatomical appearance (horn and tail size, coat type and body size) than the wild species. This is likely to have been the result of man's selection for traits which are relatively highly heritable and controlled by few genes. For example, all of the wild species have short thin tails and the tail shape differences now apparent in different breeds have arisen since domestication. In contrast, more complex traits such as behavioural characteristics are most often controlled by many genes. Although there is apparent genetic variation, there is little evidence to suggest that such traits are easily altered by selection (Goddard, 1980).

The number of *main* breeds, defined as dominant breeds of a region from which other minor breeds have been derived (Mason, 1969), have been classified according to 'tail type' in Table I.1. Clearly, the thin-tail European type dominates in terms of breed numbers but fat-tail sheep of various types are relatively widely dispersed in eastern Europe, Africa and Asia. Figure I.1(c) shows the general distribution throughout the world of the four major tail and coat types.

A division of the *main* breeds according to their major country of location (Table I.1) shows that Europe is the region with the greatest

Table I.1. The proportion (%) of the 311 *main* breeds of domestic sheep (*Ovis aries*) classified on the basis of tail-type, region of origin and usage.

Tail-type	%	Region of origin	%	Usage	%
Fat-tail	9.9	Africa	7.7	Meat	5.1
Fat-rump	2.6	Europe	51.8	Dairy	0.6
Short-tail	6.4	Middle East	7.7	Wool	10.9
Long-tail	3.2	USSR	17.7	Meat+Dairy	1.3
Short-fat-tail	2.6	China	2.2	Meat+Wool	53.7
Long-fat-tail	2.6	India/Pakistan	7.1	Dairy+Wool	3.9
Thin-tail	72.7	Australia	1.0	Meat+Dairy+Wool	24.5
		North America	2.6		
		South America	2.3		

Source: Mason (1969).

Table I.2. World sheep numbers according to region, 1989.

Region	Numbers (millions)	% of world population
Australia + New Zealand	220	18.8
Africa	163	13.9
USSR	140	11.9
Europe + United Kingdom	135	11.5
China + Mongolia	116	9.9
South America	103	8.5
India + Pakistan	100	8.4
Other	193	16.5
Total	1170	

Source: Anon. (1991).

diversity of breeds. However, an examination of the distribution of sheep numbers in the world reveals that USSR, Australia and New Zealand and China dominate total sheep numbers with over 40% of the world's sheep (Table I.2). The Merino has had a major impact on sheep populations of these countries in the last century and is also the major breed found in South America in pure or crossbred form. It has been calculated from

Table I.3. The behavioural characteristics favouring domestication.

Group structure
Large hierarchically structured groups
Males associated with females for at least part of the year

Sexual behaviour
Promiscuous mating behaviour
Sexual signalling based on movements and posture

Parent–young interactions
Precocial young establishing species bond at an early age

Responses to man
Short flight distance

Food and habitat requirements
Herbivorous and adaptability to a wide range of conditions

Source: Adapted from Hale (1962).

Mason (1969) that Merinos have contributed in some way to 86 of the 311 *main* breed types.

Uses

Sheep have traditionally been multi-purpose animals used for their fibre, pelts, meat and milk. The *main* breeds are classified in Table I.1 according to their use. Breed numbers suggest that sheep are used predominantly for meat and wool. However, it can be argued, on the basis of total sheep numbers, that the use of sheep for wool production dominates in countries with large sheep populations. This possibly explains the increasing influence of 'fine wool' breeds such as the Merino in the last century.

The multi-purpose nature of the sheep, along with their behavioural characteristics are probably the major reasons why this genus was 'chosen' as a species for domestication. Hale (1962) listed a number of traits favouring domestication (Table I.3) and it seems that sheep are ideally suited for domestication, exhibiting all traits to some degree.

Species and Breed Variation in Behaviour

In compiling this book the authors have had great difficulty in finding detailed behavioural descriptions of many of the breeds of sheep, particularly those of the fat-tail and fat-rump types. However, it is our belief that the similar origins of the domestic sheep breeds, their grazing habit, their relative ease of handling, their ability to conserve behaviour patterns in spite of environmental change and the difficulty of changing behaviour by selection may well mean that the behavioural characteristics of the breeds have remained similar to those originally exhibited by their wild progenitors. If this is so, and a comparison of the social behaviours of the existing wild species with those of domestic breeds support this contention (Price, 1984; Shackleton and Shank, 1984; Chapter 2), then the conclusions drawn throughout this book will have application to most, if not all sheep breeds.

Nevertheless, there is breed variation in some of the behavioural characteristics. This is particularly notable in the intensity of gregariousness and flocking responses and to a lesser degree in maternal and agonistic behaviours. There appears to be a continuum of gregariousness (Fig. I.2) varying from highly gregarious (Merino) to individualistic (Scottish Blackface). It is notable that the Merino is the predominant breed used in published behavioural studies. However, the continuum of behaviour has been recognized in the interpretation of the information which follows. Where possible the issue of how these differences may influence other behaviours or management have been addressed.

(a)

(b)

Fig. I.2. The dispersal of different breeds while grazing: (a) Scottish Blackface widely dispersed; (b) Romney ewes evenly dispersed even when confined to a limited area; and (c) Merino ewes closely associated when grazing in a large area.

Sheep Management and Handling

The close links between humans and sheep over a long period of history have resulted in management and husbandry patterns (Ryder, 1983) which appear surprisingly consistent across cultures. However, the reasons why these patterns have developed have not been documented and it is particularly notable that the relationship between patterns of management and the behavioural characteristics of the sheep have not been evaluated. Consequently, it is difficult to assess whether present methods of handling sheep make optimal use of our knowledge of the behaviour of this genus. For instance, which of the methods for the management of grazing sheep makes best use of the behavioural traits of the species? In the nomadic and trans-humant systems of management, sheep are normally shepherded by man sometimes with the assistance of guard dogs or trained 'leader' sheep. Where shepherding occurs, animals are often penned over night to protect them from predators. Is such a system limiting the ability of the sheep to utilize its capacity for flocking and flight as a means of protection against predation? Does the major alternative management system used in extensive grazing areas, namely the confinement of sheep within fenced fields with minimal inputs from man, allow a more complete expression of behaviours such as selectivity of diet?

In contrast to the grazing systems of management, intensive housing of sheep requires the animals to accept feed provided by humans and to adapt

(c)

to restricted space availability and to limited opportunity for the expression of their behavioural repertoire. Do such constraints have any implications to the well-being of the sheep?

The theme throughout this book has been to treat behaviour within the context of the underlying biological processes. Physiology rather than ecology is emphasized in this regard. In addition, prominence has been given to subjects important in animal production. Alimentation and reproduction, or more plainly feeding and breeding, are prime examples. These are dealt with in association with matters such as social behaviour which have a critical bearing on the husbandry of sheep. The intention is to provide an informed account of the behaviour of sheep, making use of the knowledge of centuries in the light of the scientific efforts and advances of the last 50 years. We believe this provides a basis for evaluating current management and handling procedures and for identifying gaps in knowledge. We also maintain that the next wave of improvements in the husbandry of sheep is to come from greater insights into behaviour and its intimate links with physiology.

1

Grazing Behaviour in Sheep

The process of domestication in sheep which began some 11,000 years ago has resulted in animals which are more docile and also more flexible in their adaptation to new feeds and other aspects of the environment than almost any other species of domesticated animal. By the time of the Greek ascendancy in the world there had been some 3000 generations of domesticated sheep. Michael Ryder (1983), in his book, *Sheep and Man,* has researched the history and archaeology of sheep and has summarized the writings of the Greeks in relation to sheep husbandry. Sheep were left in the care of shepherds over summer and in many areas of the world they trekked over long distances to utilize fresh mountain pastures. Sheep were kept in pens during winter and were fed barley, leaves of elm (Fig. 1.1), oak, poplar and fig, together with straw, pressed grapes after wine making, and bran. Vanro wrote that clover and alfalfa were most acceptable to sheep and both made them fat and milk well. Cato recommended that clover and beans be grown as winter crops for sheep.

In the light of this long history of domestication which now takes in around 4000 generations and the evidence that domesticated sheep still exhibit behaviours similar to those of their wild counterparts which live from the Arctic to the tropics, it can be expected that sheep have a high degree of adaptability to the environment in which they live. The successful domestication of sheep has been associated with their close relationship with people as shepherds. Within the last three centuries the shepherding system has occasionally been replaced by management systems which allow for irregular rather than consistent interference by people. Some sheep are kept in small fields containing highly nutritious grasses and others allowed to roam vast areas restricted only by the boundary fence.

Fig. 1.1. Winter fodder.

Adaptability of Grazing Behaviour

There are four wild sub-species, several feral breeds of sheep and some 800 to 1000 breeds of domestic sheep living in the world. These animals live in almost every place on earth except Antarctica and eat a wide range of shrubs, grasses, forbs and lower forms of plant life. Several examples provide an indication of the means that sheep adopt to obtain enough nutrients to survive, grow and reproduce.

The most northerly island of the Orkney group, North Ronaldsay, is

Fig. 1.2. Sheep grazing seaweed.

980 ha in area. In 1839 a large stone wall was built around the more fertile ground which deliberately confined the Orkney sheep to the outer perimeter of the island (17 km coastline). This perimeter contains 200 ha of rocks and coarse salt-tolerant grass as well as a 260 ha intertidal area. The sheep have survived by eating one of the brown seaweeds, *Laminaria* sp., which grows in the intertidal area (Fig. 1.2). Grazing times are governed by the tide with an increasing number of sheep grazing the seaweed as it becomes exposed at low tide (Fig. 1.3). The time sheep spend resting is also governed by the tide rather than the normal day/night cycle since, except for ruminating, sheep are inactive during high tide (Paterson and Coleman, 1982).

In the 1930s about 70% of sheep living in north-western Australia died during a prolonged drought. However, on one property, mortalities were reduced to 1.2% by providing an alternative feed supply. Although *Acacia aneura* leaves are edible they are not normally accessible to sheep. Since the mulga is not killed by felling the tree, it was harvested by lopping the trunk at 1.5 m and a routine was established by cutting acacia twice a day (Nichols, 1944). Sheep found the feed by responding to the sound of the axe or, these days, perhaps it would be the chain saw!

On the other side of the world, Stone's sheep live in the mountain ranges of British Columbia. In summer they eat soft, high-quality pastures of the mountains, but after the first severe frosts they move to the valley floors where they eat frozen herbs and fallen leaves from the deciduous trees. As

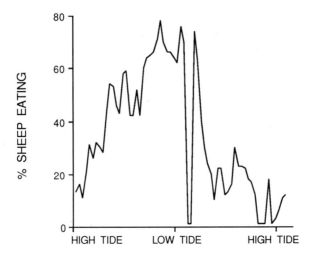

Fig. 1.3. Percentage of sheep eating seaweed at times which are related to the period around low tide (data from Paterson and Coleman, 1982).

the snow depth increases, the sheep move up the mountains where they must seek areas containing leaves of shrubs and herbs. Later in the winter they change to sedges and dry grass. By contrast, the Mountain Bighorn sheep living in Arizona forage through the year on a small variety of shrubs, forbs, trees and cacti. They break open and obtain cactus flesh by butting and kicking the plant.

Grazing Behaviour and Habitat Use

The long-term activities of sheep are associated with their requirements for food, water and rest. The latter is important for rumination, food comminution and efficient digestion. This section will examine the way sheep use the habitat they live in to obtain feed and how their activities are integrated over a 24-hour cycle.

Time spent grazing

The time a sheep spends grazing and the resultant intake is governed by a multiplicity of factors. In general terms, sheep graze for about eight to nine hours a day with the maximum time being about 13 hours when the feed supply is limited.

Sheep, living in maritime climates such as New Zealand and the United Kingdom with cool summers and having large quantities of highly digestible grass and clover, eat in feeding bouts which last from 20 to 90 minutes.

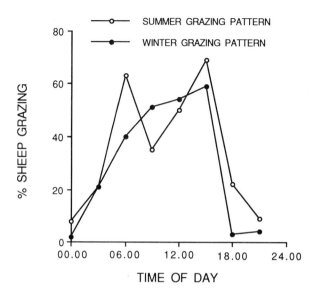

Fig. 1.4. A generalized summer and winter pattern of grazing for sheep living in continental climates with hot summers and variable-quality pasture.

As many as nine of these bouts occur over the 24-hour cycle and each is followed by a period of 45 to 90 minutes when sheep lie down and ruminate or rest. The feeding bouts cease around 2200 h and recommence about 0700 h but two 60-minute bouts of feeding can occur over this nine-hour period.

A generalized pattern of grazing activity throughout the year for sheep living in a continental environment with hot summers is shown in Fig. 1.4. Initiation and cessation of grazing activities correspond roughly to periods when there is some light in the sky. Grazing is often concentrated around the first four hours after dawn and in the last four hours around sunset but can easily start before dawn and extend long into the dark.

In pen studies, operant tests have shown that sheep prefer to spend an average of 18.5 hours in illuminated pens (Baldwin and Start, 1981). However, having light is not a high priority for sheep since they are only prepared to interrupt an infra-red beam for 4.8 minutes per 24 hours to keep a light turned on.

Habitat use

Many breeds of sheep living in undulating to steep hilly country establish a daily pattern of movement. This pattern becomes sufficiently routine with successive generations for the sheep to form what is known as a home range.

Rocky Mountain Bighorn and other wild sheep living in northern America, feral Soay sheep and domestic Scottish Blackface and Cheviot all establish home ranges. By contrast, Merino sheep do not appear to show home-range behaviour perhaps because generations of sheep do not use the same area continuously.

In hilly topography, sheep rest overnight in the higher areas then establish a pattern of movement down to the valley in the morning. The speed of movement varies. In some breeds it can be relatively unhurried and have a considerable component of grazing. At the other extreme, Norwegian sheep have a pattern of rapid intermittent grazing interspersed with trotting or running 10 to 50 m before grazing again (Warren and Mysterud, 1991). After resting in the shade at the bottom of the valley through the midday period, sheep graze slowly to the night resting site.

Sheep have been shown to congregate in areas which contained plants that are often grazed. In the Norwegian study, sheep preferred an area where particular plant species grew because of the site's high fertility and good drainage. In New Zealand the use of the higher mountainous country was changed by application of fertilizer to areas of perennial grasses growing in the valleys. The fertilizer improved the acceptability of the perennial grass to sheep which congregated on these valley areas thus diminishing their use of the mountain peaks and reducing erosion.

On flat terrain use of the habitat is less easily predicted. It is said there is south-east drift of herbivores in unfenced areas of arid Australia in response to the prevailing wind. However, wind is rarely a component determining the day-to-day direction in which sheep graze. In hot environments, long periods are spent near the watering site and the night resting area varies from one night to the next (Lynch, 1977).

Patchiness of habitat use

When habitat use by sheep is subjected to detailed study there is evidence of considerable unevenness in the use of areas. While annual stocking rate has been used as a measure of the carrying capacity of the land, the actual utilization of the land is more related to stocking intensity. This can be estimated since it has been shown that time spent in an area is proportional to the faeces deposited by the sheep as they visit that area. Lange (1985) has shown that in a 1 ha paddock of high-quality improved pasture the stocking intensity of various parts of a paddock varied from one-eighth to eight times the paddock average stocking rate (SR) of 10 sheep per ha (Fig. 1.5a). In a 2200 ha saltbush paddock which had a SR of 6.5 ha per sheep he showed that the stocking intensity of parts of the paddock varied from 0 to 800% of the paddock average (Fig. 1.5b). One-third of the paddock showed higher than average stocking intensity while two-thirds had lower than average or no use. For some reason only one of the two

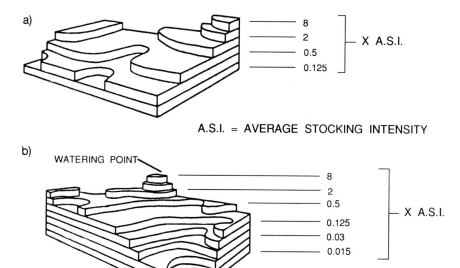

A.S.I. = AVERAGE STOCKING INTENSITY

Fig. 1.5. (a) A three-dimensional graph showing the utilization of various areas of a small (1 ha) paddock. The paddock had an overall stocking intensity ranging from 8 times to 1/8 the paddock stocking rate. (b) A three-dimensional graph showing the utilization of various areas of a large (2200 ha) paddock. The paddock had an overall stocking intensity ranging from 8 times to 1/64 the paddock stocking rate with some areas not being grazed at all (reproduced with permission from Lange, 1985).

available watering points was used and sheep grazed either towards the south-east or towards the north.

These uneven distributions of sheep in large areas of saltbush shrubland and in small highly improved paddocks suggest that this is a general phenomenon. The reasons for such variation in stocking intensity over the landscape, other than the high use of some areas related to eating preferred plants, drinking and day or night resting, remain to be determined. It is clear there are aspects of grazing behaviour of sheep related to feeding and the utilization of the habitat which require further study.

Habitat and memory for feed sites

A grazing ruminant appears to have an excellent memory of the location of preferred foods. This has rarely been studied in sheep, but recently was examined in deer which were offered three feeds in a 0.5 ha enclosure (Gillingham and Bunnell, 1989). Initially, apples were shown to be preferred to pelleted dairy food or alfalfa. These foods were placed separately

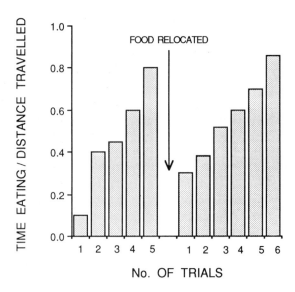

Fig. 1.6. The effect of experience of the location of feed on the efficiency of finding (time eating/distance travelled:s/m). The arrow represents the time when food was relocated which has resulted in a markedly lower efficiency (data from Gillingham and Bunnell, 1989).

in containers in a pattern to create discrete patches of food at 5 m intervals throughout an area. Deer searched the enclosure for apples. When apples became rare the second preference was accepted. Initial studies showed that the time spent eating an apple did not change over the experimental period so that animals became more efficient (measured by time spent eating per distance travelled) in finding the apples with increased experience of the apples' location. When the distribution changed they became less efficient. Having achieved an efficiency of 0.8 seconds of eating time per metre walked, the animals were provided with food in a different distribution and their efficiency dropped to 0.4 before building to 0.8 after six trials (Fig. 1.6). The first search path of the deer after relocation resembled the path taken by the last search path of the previous location. The data indicate memory of success of the previous foraging expedition.

Searching behaviour for food was also suggested by an experiment in which sheep initially grazed a paddock containing 50% of a highly preferred plant, alfalfa (Gluesing and Balph, 1980). After 10 days they were placed in another paddock containing only 4% alfalfa. Initially sheep walked 50% further while grazing, but within several days they reduced the distance walked to that observed in the first paddock.

Habitat and Water Use

Sheep have several physiological mechanisms which enable them to exist for long periods without drinking water. Occasionally the drives of hunger and thirst come into conflict and the sheep's behavioural responses demonstrate their flexibility in satisfying both drives. When sheep are grazing in small fields, the time taken for sheep to move from the grazing to the watering site and back is very short. In larger areas this is not always so.

The manner in which sheep can balance their need for water and intersperse this with grazing has been studied on two properties in the arid area of Australia. On one property where saltbush is the major plant component, the distance grazed from the water site is limited by the need to drink daily (Fig. 1.7). Little grazing occurs more than 7 km from water. During summer, sheep must drink once a day when other plants are eaten with saltbush. In autumn, saltbush is the only forage available to eat and sheep must drink twice a day (Squires, 1981). In such environments where sheep do not graze in the middle of the day when temperatures are high, the time spent walking to and from water is about one and a half hours in summer and is doubled in autumn. On another property in the same area with a different plant community and no saltbush, sheep may spend as much as two to three days grazing in summer without walking to water. It is only when the maximum temperature remains above 42°C for more than three consecutive days that sheep walk to water every day (Lynch, 1974). On both

Fig. 1.7. Sheep congregating at the water supply.

properties during winter when grasses and forbs were available no sheep, even lactating ewes, spent time walking to water.

In another study, sheep tracks around a watering point were shown mathematically to be 15° off radial, indicating precise navigational skill (Lange, 1969). Since sheep faeces were scattered over the interstices of the track network it would seem that the tracks are for walking and limited grazing and inter-track areas are used for additional grazing. This behaviour ensures that virtually every plant within the track system will be subject to grazing. There are, of course, many factors which will modify this track pattern including, trees, fence lines, heterogeneity of vegetation and topography.

Habitat and resting behaviour

In cold climates, sheep of most breeds rest at night at or very close to the highest site and near to the place where grazing ceases (Taylor *et al.*, 1987). It is not known why sheep rest at these high sites which may be as little as 1 m above the lowest site in the paddock or many metres up the side of a mountain. There has only been one set of observations relating night air temperature to resting site. This showed that the colder the temperature the higher the sheep moved up the hill to rest at night (Arnold and Dudzinski, 1978). In hot climates sheep appear to rest at night at the site where they stop grazing and are not necessarily seen at a site higher than the rest of the paddock.

During the day sheep of all breeds rest at the site where they cease grazing unless they have spent a long time walking to water. In the latter case, up to 10 hours of the day are spent resting either under shade trees or near water. If no shade is available, sheep may spend many hours resting on top of the dam wall. When the water supply is piped to a trough and there is no surface water supply, sheep can be seen with their heads under each other's belly.

Feed Intake and Grazing Behaviour

A broad picture has been presented of the way a sheep occupies each 24 hours of its life. The most dominant behaviour each day is grazing so that the sheep obtains enough food for its maintenance, growth and reproduction (see Fig. 1.9 below). The resultant food intake which is a central facet of grazing behaviour is itself composed of three behavioural variables, grazing time, bite size and bite rate. These variables, which can be modified by many plant and animal components, are shown at the centre of the flow chart (Fig. 1.8) with many of the factors which will interact with food intake shown around the periphery. The whole flow chart is clearly a two-

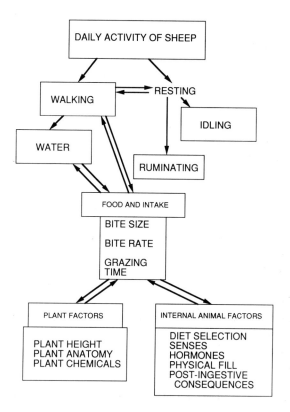

Fig. 1.8. A flow chart of the many factors impinging on grazing behaviour.

way process with one parameter affecting and being affected by one or several other parameters.

This following section will discuss the component parts of grazing and will then consider some animal and plant factors which cause variation in aspects of the grazing behaviour of sheep. Diet selection will be discussed in a separate section.

Prehension of plants

There are several mechanisms which sheep use to harvest food. By grazing very close to the ground sheep can harvest many prostrate plants. Grasses are prehended then torn when the head is moved posteriorly with a sudden jerking movement. The head may swing laterally and more food is prehended while the fore or hind leg takes one step forward. When eating shrubs the sheep can either strip the branch of leaves, break the twig and chew it, or pick off discrete leaves. Sheep will dig through soft snow to search for dead grass if there is no other food available and can survive on clover seed in what appears to be barren paddocks.

Grazing times, bite size and bite rate

During the first half of this century, grazing times were measured by direct observation (Fig. 1.9). Over the last 30 years, improved transducer and data-logging techniques have been increasingly used. More detailed study of feeding behaviour brought about the realization that grazing time was only one factor related to intake. The rate of intake (g/min) is made up of bite size and bite rate and grazing time, all of which can vary. Maximum bite size is related to the width of the mouth and the depth of the mouth cavity. Grazing time is composed of prehending, chewing and preparing the bolus to be swallowed; it also has a component of walking to a new site and searching for food and selecting.

Behavioural observations show that in both small and large flocks, sheep do not commence or cease grazing at the same time (Arnold and Dudzinski, 1978). In a large flock the time from the first to the last sheep to commence grazing can vary by up to 45 minutes (Lynch, unpublished data). The addition of readily available carbohydrate often reduces the time spent grazing, the substitution effect, while the addition of a protein-

Fig. 1.9. Two Scottish Blackface sheep grazing.

rich supplement increases voluntary intake and presumably grazing times, of a low-quality roughage pasture.

Although pregnancy, lactation, climate, shearing and the addition of a supplementary feed have been shown to affect time spent grazing, it is difficult to predict the effect of these varizbles on grazing time because the two unrelated factors of bite size and bite rate are independent. There is even a substantial level of variation in grazing time of an individual sheep from one day to another. There have been suggestions that social facilitation operates to synchronize the commencement and cessation of grazing (Tribe, 1950). If this is so, it is not at the flock level though it may occur within subgroups of the flock.

Plant factors affecting intake

The preference of sheep for various parts of the plant has a marked effect on intake per unit time (Hodgson, 1982). The following section will discuss the way sward structure affects this rate of intake.

Bite weight

In pen experiments it has been shown that plant height and density affect rate of intake. In the field there are great problems in achieving adequate experimental control over factors such as pasture height and bulk density. There has been some success with single forage species, but complex plant mixtures are far more difficult. Bite weight and hence bite volume is also affected by the density of the pasture at the site of grazing. Allden and Whittaker (1970) have related the rate of herbage intake (g/min), size of bites and number of bites and grazing times to the herbage available and length of tillers in Fig. 1.10. Bite size and rate increase with increase in the pasture height and tiller length, while bite frequency declines when tillers are longer than 5 cm or there is more than 1000 kg/ha of pasture.

Sward structure

As a generalization, sheep which have good-quality pasture available eat that which can be eaten most rapidly; resistance to harvesting is a major factor affecting rate of feeding (Kenney and Black, 1984). Clovers, which are far easier to prehend, bite and chew, are eaten more rapidly than grasses.

In a recent study, a range of ryegrass swards were created by maintaining pastures at heights ranging from 30–120 mm for almost six months (Penning *et al.*, 1991). The aim of this work was to relate the management of continuously stocked pastures to animal production from these pastures. The sheep were automatically monitored for herbage intake, time spent

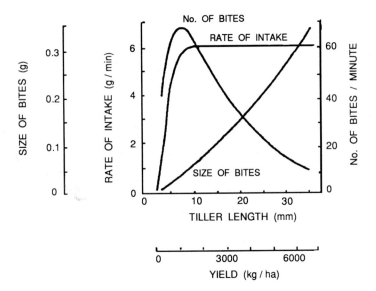

Fig. 1.10. The effect of herbage available (kg/ha) and tiller length (cm) on the number of bites (per min), size of bites (g) and rate of intake (g/min) (reproduced with permission from Allden and Whittaker, 1970).

grazing, resting and ruminating, prehension and mastication rate. The relation of these parameters to herbage height can be seen in Fig. 1.11. Grazing time and prehension rate decreases with increasing height while mastication and rumination rate increases. Jaw movements of sheep in all pastures were constant at 150 per minute with the rate of prehension varying from 20,000–60,000 bites per day. This high-quality pasture had an optimum height between 30 and 60 mm which maximized lamb growth rate and minimized change in ewe live weight.

Sheep can exhibit preferences about the depth they graze into the pasture. When placed on a high-quality ryegrass pasture, sheep do not graze the area dominated by pseudostem and dead material (Fig. 1.12), but if the pasture is mixed and contains white clover sheep will graze the pseudostems. The plant/animal interface is an important but difficult area of study. Grazing can affect plant regrowth which can result in subsequent food supply as well as the grazing behaviour of the sheep being altered.

Diet Selection

Diet selection is based on sheep making a decision concerning what plant species, individual plants and parts of plants it will eat. Sheep have a relatively narrow muzzle so they are capable of selecting plant parts with great

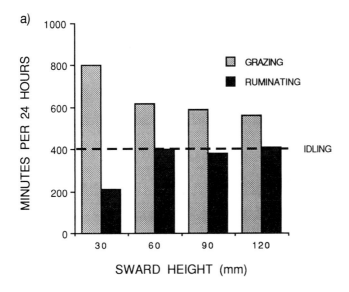

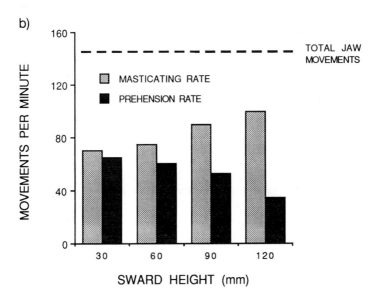

Fig. 1.11. (a) An example of the effect of pasture surface height on time spent grazing, ruminating and idling per day and (b) on total jaw movements, prehension rate and masticating rate (data from Penning *et al.*, 1991).

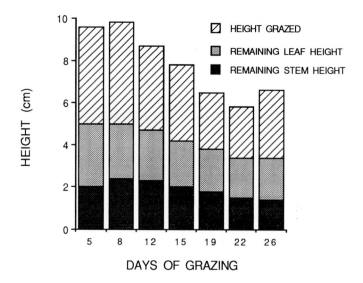

Fig. 1.12. The depth that sheep graze a 50 mm ryegrass pasture over a month while leaving the stems untouched (data from Barthram, 1980).

precision. The preferences sheep exhibit may increase or decrease the rate of eating. Hence, the rate of eating is lowered when sheep select only green material from a pasture which is mainly dead. Intake per day can be maintained through a greatly increased grazing time.

Examples of diet selection

Comparisons of various methods for determining diet selection indicate that collection of extrusa from a fistulated oesophagus of sheep is the most accurate provided the sheep are properly cared for and managed the same way as other sheep in the flock. The disadvantages are that this method is very time consuming in the maintenance of animals and in the identification of plant parts that are collected from the extrusa (Laycock *et al.*, 1972). Fistulated and normal sheep which are managed similarly select diets of the same botanical composition (Jung and Koong, 1985; Forbes and Beattie, 1987).

The highly selective nature of sheep when offered a high-quality grass/ clover pasture is shown in Figs 1.13a and b in which the pasture species available are contrasted with that selected. The overwhelming selection for leaf is evident and there was little evidence for selection of clover, pseudo-stem or dead material. The sheep grazed within 0 to 4 cm from the ground where 95% of the green leaf was located even though they had to penetrate the surface canopy where there was abundant dead material up to 20 cm

(a)

SWARD

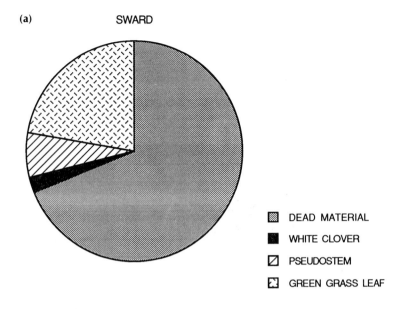

☷ DEAD MATERIAL

■ WHITE CLOVER

▨ PSEUDOSTEM

⚃ GREEN GRASS LEAF

(b)

OESOPHAGEAL BOLUSES

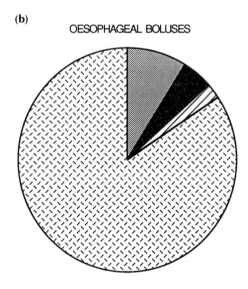

Fig. 1.13. Botanical composition of (a) the pasture, and (b) the oesophageal extrusa (data from L'Huillier *et al.*, 1984).

The Behaviour of Sheep

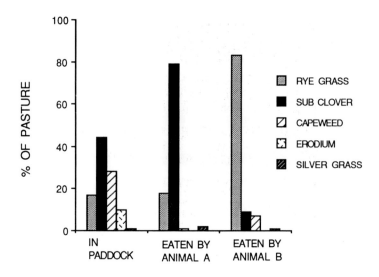

Fig. 1.14. Dry weight percentage of five plant species in a pasture and in the oesophageal extrusa of two sheep (reproduced with permission from Broom and Arnold, 1986).

high. In all other pastures in this experiment the leaf was at the surface of the pasture and was selected.

In another study (Broom and Arnold, 1986), sheep grazed a pasture containing ryegrass, subterranean clover and a species of *Erodium, Vulpia* (silver grass) and *Arctotheca* (capeweed). There was insufficient herbage available to maintain live weight until week 6 of the study and yet, for some reason, *Erodium* sp. was not eaten and very little capeweed was selected although both species are known to be edible and in greater abundance than the diet selected (Fig. 1.14). There was a big variation in the pasture preferences between sheep when the oesophageal fistula samples were taken. These results illustrate several exceptions to any generalizations concerning factors affecting food intake: (i) sheep do not necessarily graze species which can provide the most rapid rate of food intake; (ii) sheep do not necessarily graze the taller pasture or most readily available plant material; (iii) sheep do not necessarily forage optimally in relation to energy gains.

The sheep's selectivity for herbage which has high digestibility and nitrogen concentration when compared with the herbage on offer is seen in Fig. 1.15. The sheep is probably not selecting for either nitrogen levels or high digestibility, but it is not understood what parameters are being selected which are correlated with nitrogen and digestibility.

Other research on diet selection has shown that sheep can select a single species as 80% of their diet during a period when the species was present as

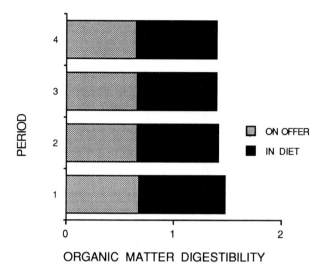

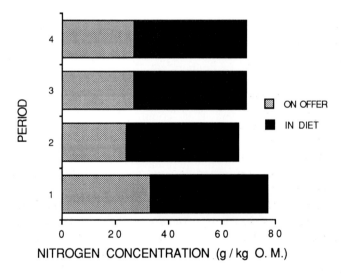

Fig. 1.15. The selectivity of sheep for high organic matter digestibility and high nitrogen concentration compared with that availalbe from pastures (data from Curll *et al.*, 1985).

1% of the mass of herbage available (Leigh and Mulham, 1966). This example of high selectivity for green material was in a period of the year when shrubs or dry material were readily available to sheep.

Factors affecting diet selection

Consistent differences have been found between Corriedale, Dorset Horn and Merino breeds in the preference these breeds have for various plant species (Arnold *et al.*, 1981). Differences also exist in the diet selected by individuals within the same breed.

Level of hunger can also have an effect on diet selection. There is evidence that as hunger is partially satiated by eating a high-quality feed an increased selectivity for quality of diet occurs (Jung and Koong, 1985).

Experience

Prior experience has a marked effect on determining whether sheep will eat a food. The experience can come from several sources.

1. *Mother* The lamb spends the period, until weaning, grazing close by its mother. Unweaned domestic or feral sheep have their progeny near them for much longer; the association lasts at least three years.

2. *Other adults* Although the lamb may be permanently separated from its mother, either accidentally or deliberately, there are management systems whereby young ewes and wethers graze with older animals either of the same or opposite sex.

3. *Peers* Grazing with peers is a common management system for domestic sheep.

4. *Previous history* The other source of experience relates to the previous grazing history of the sheep. They may have been grazing in different environments or on markedly different plant species, for example, shrubs rather than pastures.

This section will discuss what is known of the way experience affects acceptance and rate of intake of plants and feed supplements such as grains, pellets and blocks. In some studies, plants are the appropriate unit while studies on novel foods and the process of familiarization are more easily done with supplements.

Foraging selection and skills

Information on the diet selected by sheep moved to an environment with very different pasture species is very limited even though this is a common management procedure. Sheep which were moved from dry areas

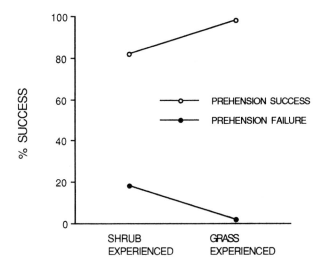

Fig. 1.16. Percentage prehension success and failure of shrub- and grass-fed lambs when offered crested wheatgrass (data from Flores *et al.*, 1989b).

containing native pastures to improved pastures in coastal Australia showed quite different pasture preferences to those sheep raised on improved pastures. These preferences changed slowly over 6 to 12 months.

In one experiment, sheep were moved from two separate native range-land pastures to the high-quality pastures of humid New South Wales and compared with sheep reared from birth on improved pastures (Langlands, 1969). The latter consistently selected a diet with a nitrogen content which was significantly higher than the diet selected by the two groups reared on rangelands but grazing the good-quality pastures for more than three months.

Sheep reared on grasses had a higher bite rate per minute and a larger bite size of both the vegetative and flowering stage of the grass than sheep experienced in harvesting shrubs rather than grasses. The experienced grass eaters showed more dexterity at prehending flowering stems and heads of grasses than the shrub eaters (Fig. 1.16). Similarly, sheep experienced in foraging from shrubs did so more efficiently than those animals experienced only in eating grass. This efficiency was related to rate of intake and the success of obtaining food by stripping leaves, plucking individual leaves and breaking twigs (Flores *et al.*, 1989a,b).

Single exposure to food and results of familiarity

When naive sheep were exposed to wheat in a trough for one 15-minute period (single exposure) there was no intake of wheat the next time they

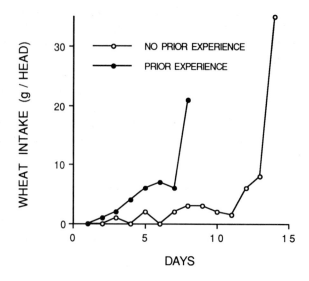

Fig. 1.17. Daily wheat intake (g/head/day) of sheep which were familiar with eating from a feed trough but not eating wheat, and sheep which had never previously seen the feed trough or wheat.

were offered it. After repeated exposure there is an effect of familiarity. In studies on acceptance of wheat by Merino sheep, a 15-minute daily exposure of individual sheep to the grain was routinely used (Chapple and Lynch, 1986). Thirteen days were taken before the sheep ate even 10 g of wheat in 15 minutes, but the quantity of grain consumed per day increased very rapidly. It was found that the time taken for sheep to eat wheat could be reduced by half if the sheep had previously been fed hay from the troughs. Obviously this overcame some of the neophobia (fear of new things) experienced by the sheep (Fig. 1.17).

Age

Behavioural studies have shown that young sheep accept new foods more readily than adults. Pre-weaning exposure to molasses–urea blocks had a greater influence upon later acceptance of the blocks than post-weaning exposure. In fact, exposure to molasses–urea during the first nine months of life resulted in a greater likelihood of animals eating the supplement later in life than if animals were first offered the supplements after nine months of age (Lobato *et al.*, 1980). When sheep were kept in age groups and offered oats, weaners had a lower percentage of non-feeders and required fewer days to start feeding, compared with older sheep.

In an experiment to determine at what age lambs could be exposed to wheat and eat it later, groups of lambs with their mothers were offered wheat for 15 minutes per day over five consecutive days during weeks 1, 2, 3, 4, 5, 6 or 7 of life (Chapple and Lynch, 1986). Subsequently, the lambs which were more than three weeks old when first offered wheat (Fig. 1.18) ate an average of 120 g/head when offered the food after weaning. This knowledge appears to remain with the sheep for life and to still be used after years of encountering no wheat. Sheep younger than three weeks when first offered wheat ate virtually no wheat when tested. These differences in intake that occur as a function of age have been confirmed and reviewed by Provenza *et al.* (1992).

Social transmission of feeding behaviour (social models)

Learning to eat new foods may be greatly hastened if animals watch and participate with feeding conspecifics performing the task. Social transmission of information is a more rapid process than laborious trial and error learning. The importance of a social model in influencing the plants selected by lambs has been well established.

In a cross-fostering experiment, Key and MacIver (1980) showed that Clun Forest and Welsh Mountain lambs preferred the distinctly different improved pasture or tussock and heather eaten by their foster mothers. The authors state that 'it would appear ... that sheep are not born with innate behavioural patterns determining their grazing habits but rather that the latter are acquired by copying the habits of their natural or foster mother'. It is almost inevitable that the mother is the social model since from six weeks of age until mother/young separation the ewe's lamb is always less than two metres from her (Hinch *et al.*, 1987).

Young lambs which see their mothers eating grain have a life-time memory for the feed and have been shown to eat it readily some three years later (Fig. 1.19). It is likely that sheep which experience eating particular plants will remember them for years also. This long-term memory makes studies on animal and plant factors affecting intake and diet selection very inaccurate unless the previous feeding experiences of the sheep are known.

The mother and its lamb are always an isolated unit in experiments on social transmission of feeding. Very often, lambs are part of a flock and could be influenced by adults other than the lamb's mother. In one experiment lambs one or seven weeks old were offered wheat either with their mothers or with adult wethers experienced in eating wheat. The results showed that provided the lambs were seven weeks old, they could learn from any experienced adult (Fig 1.20). In another experiment lambs exposed with their mothers subsequently ate far more grain and shrub leaves than those exposed with a dry ewe (Thorhallsdottir *et al.*, 1990), but lambs exposed with a dry ewe also ate more than naive lambs. More

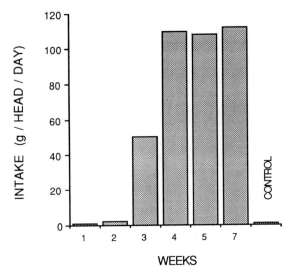

Fig. 1.18. Mean wheat intake (g/head/day) after weaning of groups of sheep offered wheat at 1, 2, 3, 4, 5 or 7 weeks of age with mothers experienced in eating wheat. Control sheep saw wheat for the first time after weaning. Sheep offered wheat in week 3 ate significantly less than those from weeks 4 to 7, but more than the other groups.

importantly, the experiment showed conclusively that the lambs had to participate with their conspecifics rather than watching them to ensure they learnt to eat the novel feeds. Other data (Chapple *et al.*, 1987) showed that if experienced wheat eaters were in the centre of groups of four sheep naive to wheat, the latter rapidly learnt to eat the wheat (Fig. 1.21). In this case the naive sheep could watch the experienced eater and could eat from their own trough, which is in contrast to the experiment of Thorhallsdottir *et al.* (1990), where no trough was present.

Role of senses

The important role the senses play in the selection of diet has often been avoided by animal research workers. It is difficult research, particularly because of the interactive relationships of taste and smell and the role of prior experience and memory. These factors may have confounded the results of some experiments. Most research on the role of the senses has been done by sectioning the nerve supply to the senses or the application of osmic acid to kill the nerve endings. However, the technique of using surgery or chemicals to destroy the sense of sight, smell or hearing may be of little benefit in trying to assess the relative importance of various senses in diet selection since the adaptation to loss of a sense can be very rapid.

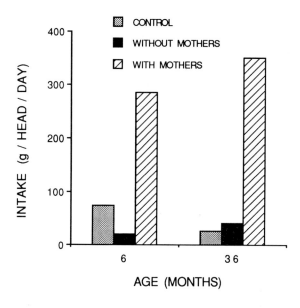

Fig. 1.19. Mean wheat intake (g/head/day) of groups of sheep offered wheat at 6 and 36 months after having been exposed to wheat pre-weaning with or without mothers. Controls had never previously seen wheat until the week of testing.

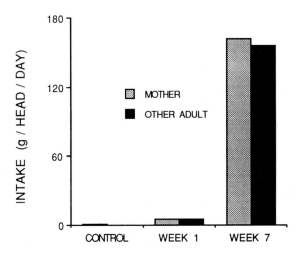

Fig. 1.20. Mean wheat intake (g/head/day) after weaning of groups of sheep offered wheat with their mothers or other adult sheep at age one or seven weeks. The control group saw wheat for the first time after weaning.

OBSERVATIONAL LEARNING IN PENS

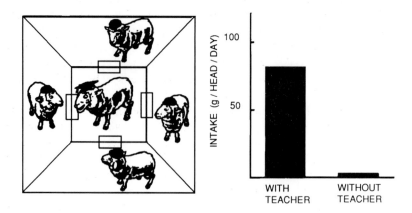

Fig. 1.21. Sheep which had never previously seen wheat learned to eat wheat by watching an experienced wheat eater. The wheat intake (g/head/day) is the mean intake over five days and is compared with the intake of groups which had non-eaters in the centre pen. All sheep had individual troughs.

Olfaction

In the literature there are differences of opinion concerning the importance of the sense of olfaction in diet selection. Sometimes there appears to be a difficulty in interpreting data on the behaviour of sheep with or without one or more of the species' senses. If there is no difference in diet selection by sheep with the sense of smell permanently impaired, smell is either absolutely unimportant to the process of selection or sheep are able to adapt in its absence and to replace it with other senses.

Evidence for the importance of olfaction in diet selection is provided from an experiment in which five-week-old lambs were exposed to wheat in the company of adults experienced in eating the grain. When the sheep were offered wheat post-weaning, in visually different forms as grain, flour and pellets, they ate similar quantities of each (160 g/h per day) (Chapple and Lynch, 1986). Sheep which ate the wheat grain immediately took six days before they ate barley, a seed visually similar to wheat (Mottershead *et al.*, 1985). Clearly sheep did not initially know barley as a feed by using the senses of smell or vision. Since sheep recognized immediately the three visually distinct forms of wheat and they ate them it appears that the common factor in recognition was olfaction.

Further, an interpretation of the following experiment (Lynch, unpublished data) is consistent with the importance of olfaction to feeding behaviour. Sheep were observed to select only phalaris from a mixed phalaris, ryegrass, fescue and clover pasture. After having been made anosmic (insensitive to smell) by treating the nasal mucosa with an local anaesthetic aerosol, it took sheep just 15 minutes to recommence selecting phalaris although the anaesthetic did not wear off for two hours. The behavioural sequence was as follows. The sheep stopped eating. It then prehended a small piece of phalaris leaf which it chewed. The speed of harvesting gradually increased until the sheep was selecting phalaris just as rapidly 15 minutes after anosmia was created as in the pre-treatment period. Prior to the anaesthetic, the plant may have been selected on vision initially and its selection confirmed by odour prior to prehension. After the anaesthetic, the sheep apparently selected phalaris from the mixed pasture by taste or texture. Then the sheep presumably used vision and perhaps touch of the plant on the muzzle in selecting the phalaris.

A further illustration of the role offaction has on diet selection is shown by an experiment in which sheep were offered alfalfa pellets in bins which contained various faecal odours (Pfister *et al.*, 1990). Of the feed contaminated by odours of coyote, fox or cougar, 5% was eaten compared with 95% of the pellets uncontaminated by odour (Table 1.1). There was no evidence of habituation to the odours. Sheep not only avoided the contaminated feed but also rested over the three days in the area as far away as possible from the site of the contaminated bins, which was possibly an indication of the aversion to the odours.

It seems that olfaction is important in a sheep recognizing food and perhaps selecting it from a mixed pasture. Impairment of olfaction, however, results in sheep adapting by using other senses (Chapple *et al.*, 1987).

Table 1.1. Mean feed intake over 72 hours of normal feed and feeds permeated by faecal odours.

| | Consumption (kg/head) | |
	Normal feed	Feed with odour
Coyote	8.38	0.45
Bear	5.26	4.01
Fox	8.69	0.37
Cougar	9.32	0.47
Oil of wintergreen	5.48	3.80

Source: Pfister *et al.* (1990).

Sight

The role of sight may well be to assist sheep in determining what is about to be eaten. Vision could also be an important factor in sheep deciding how far to walk before recommencing grazing. Plotting the vision of sheep at various angles from the front of the head to the back and dorsoventrally showed no vision for 2–3 cm directly in front of the nose. There is an area of binocular vision which is some 40° wide and from 30° below the vertical to 30° below the horizontal. Lateral to this area of binocular vision is monocular vision which gives 145° vision for both the left and right eyes. This means there is a 70° sector to the back of the animal with no vision unless the sheep moves its head (see Figure 2.3, Chapter 2).

Touch

Although it has not been studied, the sense of touch around the muzzle area of the sheep's head may be important. With experience, sheep may use touch for recognition of an individual plant or plant part. Since there is a blind area of some 3 cm directly in front of the nose it would be logical to expect that the selection of a particular plant part is based on odour and on touch with ingestion or rejection arbitrated by taste and/or feel within the mouth.

Effect of deprivation of senses

There have been a few studies examining the effects of depriving sheep of one or several senses on their acceptance of a new feed (Chapple *et al.*,

Table 1.2. The number (total of 16 per group) of sensory impaired sheep which accepted wheat over 5 days, and the mean wheat intake (grams per head) for five days.

Treatment	Number of animals feeding by day 5	Mean wheat intake/head (g)
Blind	14	460
Deaf and anosmic	13	362
Deaf	14	316
Deaf and blind	9	305
Control	15	198
Deaf, blind and anosmic	6	177
Anosmic	14	165
Anosmic and blind	9	141

1987). In one such study, sheep experienced in eating wheat were placed in close proximity to naive lambs which were divided into several groups which had one or more senses deprived as listed in Table 1.2. Sheep in all groups were eating wheat by day 3 and while the variation in intake between groups was large, their consumption was in marked contrast to the group exposed to wheat with no experienced wheat eater nearby. In this case no animal ate wheat. Most sheep adapted, in 15 to 30 minutes, to the loss of one or even the three senses. No one sense was more important than another in a sheep learning to eat a new food.

Preference and Palatability

Since palatability is not able to be measured in sheep, there seems little point in using the term. An increase or decrease in intake can be measured and that, together with determining the controlling variables, is more useful than the abstract concept of palatability.

It is possible to measure preference, but it applies to the particular situation under which the plant materials were tested. What sheep eat in one circumstance may be ignored in another. Clearly, as drought or other environmental factors intervene to affect pasture availability, previously rejected plants or plant components will be eaten. This suggests that small differences in palatability seen in cafeteria-type grazing tests may have little significance for animal production. There are, of course, plants which will never be eaten by sheep.

Table 1.3. The effect of various concentrations of sweet (sucrose), sour (HCl), salt (NaCl), bitter (urea) and umani (monosodium glutamate, MSG) on intake of lucerne pellets.

| | Concentration | | |
	Low	Medium	High
Sucrose	b	b	b
HCl	c	c	b
NaCl	c	c	c
Urea	b	b	b
MSG	a	a	a

a = Increased intake compared to control.
b = Decreased intake compared to control.
c = Similar intake compared to control.
Source: Grovum and Chapman (1988).

One experiment has successfully separated preference from post-ingestive effects (Grovum and Chapman, 1988). Sheep with oesophageal fistulae were sham-fed lucerne pellets which had various levels of chemicals representing sweet (sucrose), sour (hydrochloric acid), salt (sodium chloride), bitter (urea) and umani (monosodium glutamate, MSG). Sucrose and urea depressed intake while sodium chloride and MSG increased it (Table 1.3). When sucrose was fed to normal sheep (no fistula) intake was increased, while sodium chloride depressed intake. Clearly post-ingestive effects of these two chemicals were having a marked effect on food intake which is different to that seen in the palatability or sham-feeding studies.

A basis for selection of a food item

An extensive review of this subject has been provided by Provenza *et al.* (1992). The review helps an understanding both of diet selection and of why sheep eat poisonous plants.

According to general psychology, the affective (automatic and internal) and the cognitive (external and voluntary) systems process information about food. Cognitive processing in higher cortical centres with its connections through the limbic system to the hypothalmus results in odour, sight and touch being integrated with food and its taste.

The affective processing system integrates taste and the post-ingestive consequences via the brain stem then through the limbic system to the hypothalmus. The affective and cognitive systems enable food recognition and an assessment of palatability which results in diet selection (Fig. 1.22). The two different systems, together with the many other factors outlined in this chapter, provide the animal an opportunity to assess and select the food on the basis of feedback from toxins and perhaps nutrients. The consequences of having eaten a food can be positive or negative. The former will result in an increase in intake (i.e. conditioned food preferences) and may well be at the basis of the maintenance of the protein to energy balance which exists in such diverse species as rats (Tepper and Kanarek, 1989), pigs (Kyriazakis *et al.*, 1990) and chickens (Cumming, 1989). There is even some evidence that three-month-old lambs can form preferences for non-nutritive flavours paired with energy (Burritt and Provenza, 1991). Negative post-ingestive consequences result in a sheep associating a food it has eaten with a poison which stimulates the nausea centre and causes the food to be avoided or eaten in small amounts (Thorhallsdottir *et al.*, 1987). It is not known whether other forms of positive post-ingestional consequences or conditioned food preferences exist in ruminants. It is an area of research which has received little attention.

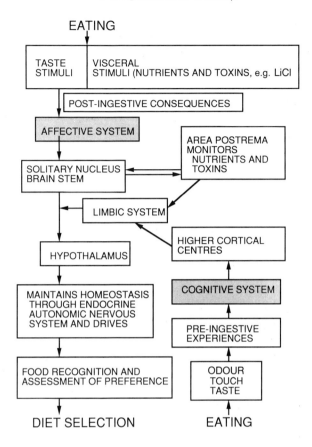

Fig. 1.22. A model of the way the affective and cognitive systems integrate with food recognition and an assessment of palatability which results in diet selection.

Conditioned feed aversions

The selection of particular plants by herbivores has a pre- and post-ingestional phase. Research workers have rarely examined the effects on future diet selection of a plant which has just been swallowed. One method of studying post-ingestional consequences is to pair the food with lithium chloride (LiCl), a gastrointestinal poison, and then study the importance of experienced, social models when either the naive animal or its social model has been given food paired with LiCl. Lithium chloride has a long history of being studied as a means of averting rats to various feeds. In humans the effects are dose-dependent and range from slight discomfort through varying levels of nausea to diarrhoea and vomiting. Provenza and his colleagues have pioneered research using paired LiCl and food to induce nausea and create conditioned feed aversion (CFA) in sheep. This work was recently reviewed by Provenza *et al.* (1992). These CFAs provide a greater understanding of the various social and physical factors which affect diet selection of sheep. CFA has also been used to explore the possibility of stopping sheep eating a particular shrub either because of toxicity to the sheep or because it is too edible.

In studies on CFA, sheep are given LiCl either in the feed, by injection or orally by capsule after a food has been eaten. An example of the experimental procedure is to give sheep pellets of a palatable material containing 2% of either NaCl or LiCl. Both salts are similarly salty. The intake of rolled barley containing LiCl dropped from 200 g to 100 g in two days. The memory of the food containing LiCl was shown by the low intake recorded two months later (Fig. 1.23). The degree of CFA was proportional with dose rate of LiCl with marked aversion to feed occurring with 150 mg LiCl/kg body weight and just one dose was sufficient to induce aversion. In another experiment, lambs were offered oats with LiCl for 12 days. The reduction in intake caused by LiCl is not absolute and depends entirely on the concentration of LiCl in the food. In an experiment lasting 12 days, the intake of sheep, which were offered *ad libitum* barley containing LiCl, stabilized on day 3 so that they were receiving LiCl at the rate of 40 mg/kg live weight, a dose which presumably is discernible but is not high enough to cause food aversion (du Toit *et al.*, 1991).

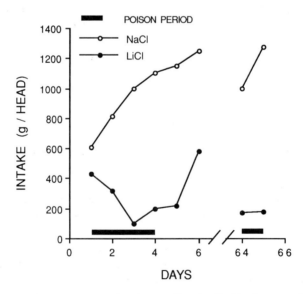

Fig. 1.23. Mean amount (g) eaten by sheep offered rolled barley containing sodium chloride or lithium chloride given during initial exposure (days 1 to 6) and on days 64 to 65 (data from Thorhallsdottir *et al.*, 1987).

Sheep are conservative feeders in that they sample a new feed and if there are no unpleasant consequences the intake of that food will gradually increase. The inherent conservatism in feeding behaviour is seen when sheep select a known rather than a novel food when given a pair choice. In such tests sheep have been shown to: (i) eat a familiar safe food and reject a novel food containing LiCl; (ii) reject a novel safe food and a novel harmful food; (iii) eat a familiar food containing LiCl and reject a novel safe food (Burritt and Provenza, 1989).

Research related to social models, feed intake and aversion has shown the mothers' greater influence on food intake of lambs compared with dry ewes or conspecifics. When mothers were averted to a food, younger lambs developed stronger aversion to foods than older lambs (Mirza and Provenza, 1991). After an aversion was created it was diminished, but not lost, if the sheep subsequently fed with non-averted conspecifics. However, the CFA was rapidly re-established by incorporating a dose of LiCl in the feed.

Conditioned feed aversions to edible and less-edible shrubs can be created in the same way as with grain and pellet foods (Burritt and Provenza, 1990). When mothers previously averted to a particular food were exposed to it with their lambs, the lambs ate less after weaning than those lambs which were alone when the harmful food was present. The non-averted social model also teaches the lamb to eat foods which the lamb had previously been conditioned not to eat. These results present further evidence that mothers may have considerable impact on the diet selected by lambs by teaching the young not only what to eat but also what to avoid.

Internal regulation of food intake

Figure 1.24 is a schematic diagram of the internal factors which have been shown to regulate feed intake, and which are extremely complex. The overall complexity is quite apparent, particularly when it is realized that this control is operating, together with the external factors shown in Fig. 1.11 mainly on stocking rate, herbage availability, grazing times, bite size and bite rate.

The overall determinant of feeding behaviour is not known and the relative contribution of individual factors which have been shown to regulate feeding behaviour remains to be determined. It has been suggested that there are short-term stimuli which control bouts of meal eating and long-term stimuli which regulate overall feed intake and hence body weight (Weston and Poppi, 1987). When meal eating occurs there are clearly many intrinsic factors producing stimuli, the net result of which is satiety or hunger. The short-term stimuli arising from the chemical and physical properties of the food operate to stop or start bouts of meal eating through stimuli in the oropharyngeal area to stimulate food intake, while gut

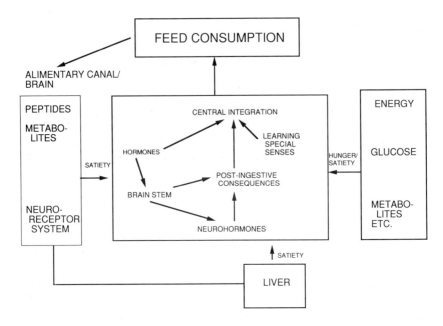

Fig. 1.24. A model of animal factors affecting food intake.

distention, gastrointestinal and liver stimuli act to inhibit feeding. The role of humoral factors is still uncertain, but exogenous cholecystokinin (CCK) decreases food intake and is probably involved in eliciting satiety (Baile and McLaughlin, 1987). At least in rats, endogenous CCK, released from the upper part of the intestine, induced satiety by stimulating receptors in the brain. In sheep, it has been shown that the longer the fasting period the greater the amount of CCK needed to suppress feeding. Also, neutralization of CCK by antisera injected into the cerebrospinal fluid delayed satiety in sheep. Since CCK changes gut motility and affects secretion of insulin and glucagon, satiety could result directly from suppression of eating behaviour or indirectly from hormonally mediated metabolic changes. Endogenous opiates have been shown to stimulate hunger-reducing activities when administered centrally. Certainly naloxone, an opiate agonist, has a marked effect on oral activities of sheep causing the mouthing and chewing of wool and even the chewing of attendant's clothing.

There is even less known about the long-term control of feed intake. It would appear that the overall 'decision' about nutrient requirements has to be the integration of multiple inputs. Kaupfermann (1985) suggests that scientists tend to believe 'exteroceptive sensory systems as being capable of more complex information processing than our interoceptive systems'. However, it is possible that these internal sensory systems are capable of information integration and processing which set levels for long-term control of feed intake.

Theories of diet selection

It is not the intention of this section to do more than summarize various theories about diet selection. Fuller understanding of diet selection will need an integrated approach rather than relying on a single discipline. The system for diet selection based on the central processing shown in Fig. 1.24 may apply generally to mammals, but with modifications for carnivores, omnivores, herbivores and ruminants.

Provenza and Balph (1990) have analysed five theories of diet selection drawing on literature associated with animal and plant science, psychology and wildlife ecology. These theories which are not mutually exclusive are: (i) nutritional wisdom; (ii) hedyphagia; (iii) morphophysiology and size; (iv) optimal foraging; and (v) learning by consequence. Some theories appear less likely than others. Euphagia or what is known colloquially as 'nutritional wisdom' suggests that animals have the innate ability to select specific nutrients. Hedyphagia or the selection of a diet 'pleasing' to the special senses and avoiding what is not is based on the concept that through the evolutionary process, what is pleasing will be nutritious. Neither euphagia nor hedyphagia takes post-ingestive consequences into account. Morphophysiology and size of species may result in ruminants ingesting herbage that differs in physical and chemical characteristics either because of body size and metabolic requirements or because they have evolved in environments with plants of vastly different digestibility. This does not cast any light on the individual variability in diet selection which is partially genetic and partially experiential. Optimal foraging theory has similar problems related to individuality. Learning by consequences, a theme of this chapter, is based on positive and negative post-ingestional consequences and experience which may be either social or have individual trial and error experiences. Of the five theories examined, learning by consequences appears to be the broadest though some of the other theories may be involved.

Innate behaviour and diet selection

The scant information on the initial responses of lambs to food has yet to be reconciled with the observation that previous experience plays a large role in the subsequent selection of a sheep's diet. Just as the newborn lamb does not know that the large, bleating, licking object is its mother who must provide milk, the newborn lamb will not know that herbage has to be eaten. In Chapter 5 the innate responses of the lamb are described whereby the lamb finds the teat and sucks within an hour of birth. It is possible that

Table 1.4. Diet selection in sheep.

---

1. Treat new foods with caution

2. Learn quickly to eat or reject particular foods and ingest small quantities

3. Ingest a number of different staple foods and simultaneously sample other food

4. Preferentially eat familiar foods for as long as possible

5. Minimize feeding on plants containing phytotoxins

6. Have searching strategy, learn where foods are and remember them

---

Source: Freedland and Janzen (1974).

the lamb has a series of innate responses which result in its eating grass, but these have never been studied. Certainly, lambs raised on artificial milk for 10 weeks did not know how to eat grass for the first half hour after being released from the pen. The initial diet selected by young lambs and the way in which this changes with experience have never been examined.

The conservative behaviour of the sheep grazing in extensive areas results in them eating a variety of plants. A good basis for a sheep selecting its diet and not being poisoned by plants has been suggested by Freedland and Janzen (1974) for generalist herbivores, but can be expected to apply to sheep (Table 1.4). Certainly sheep show the attributes related to some of these categories which are mentioned in the table. This list provides the impetus to further investigate diet selection in sheep.

Grazing Management

The pasture structure which will optimize animal production is gradually being defined. One such structure consists of a high quality dense legume grass pasture which is green and between 30 to 50 mm high. This may be achievable in the warmer seasons of the maritime climates of the United Kingdom and New Zealand, but most of the sheep in the world live in continental climates where rainfall is either irregular or insufficient to produce a long growing season. We have yet to explore the plant/shrub management systems which will optimize animal production in most areas of the world.

The conservative feeding behaviour of the sheep which enables it to survive by not being poisoned by plants mitigates against ready acceptance of supplementary feed designed to medicate, increase growth rate or fatten sheep.

Feed supplements

There are many situations when sheep must be given feed other than living plants so that they can be kept alive or their productivity increased. The supplement can take the form of feed blocks, grains, fluids or extruded feeds which contain energy, protein or minerals. When sheep are offered feed supplements, other than good-quality hay, there is a large and unpredictable variation (10–100%) in the percentage of animals which will eat them (Juwarini *et al.*, 1981). When all sheep have learnt to eat the supplement there is still a three- to five-fold variation in intake between animals.

The reasons for failure to eat the supplement are known to include lack of experience with novel food or novel container holding the food (neophobia) and the inedible nature of the novel food. Once sheep are trained to eat, some sheep are still seen standing within 3 m of a supplement but not venturing in to eat it. Extremes of age is another factor since old and young sheep are often seen standing near the supplement but failing to eat. Overt dominance from other sheep has been shown to account for 25% of the observations when sheep failed to eat the supplement. It is possible that more subtle forms of dominance or subordination exist and are shown by small changes in recognition or in orientation of the head or body of one animal to another. Other more obvious behaviours have been explored with little success.

Fig. 1.25. Feed has been placed in a large diameter circle. It has been shown that as the dams eat the feed the lambs are learning and will readily eat it even years later.

If ewes eat the supplement, the simple strategy of feeding ewes and their lambs a minimum amount of grain in the last week before weaning can produce the situation where all sheep will eat that particular grain (Fig. 1.25). Social transmission of feeding behaviour ensures that this learned behaviour to eat grain is transmitted from one generation to another in perpetuity. One limitation to this management strategy is that the same grain must be used each year. In this connection, we have shown that sheep that eat one grain type must learn to accept another.

Other possible methods of ensuring that supplements will be eaten include the pairing of an accepted supplement with an odour and using that salient odour with other supplements. The large variation in intake of supplement by animals which know how to eat it also needs urgent attention.

The role undernutrition plays in the speed of sheep learning to eat new foods has never been investigated. It may well be important in any management procedure when it is necessary to feed sheep with novel substances.

Conclusions

The comments on grazing behaviour of ruminants made by J.D. Ivins some 40 years ago are just as valid now. He was distressed about the variability in diet selection and stated:

> It is therefore submitted that conclusions based on the behaviour of a small number of animals, or over short periods, can be quite misleading and often erroneous and cannot be accepted in their entirety. Difficult and dangerous as it is to dogmatise on any particular biological aspect it is even more dangerous to dogmatise on the subject of animal behaviour. Categorical claims concerning the behaviour of grazing animals must be viewed only in relation to the particular circumstances, for these observations show quite clearly that animal behaviour is not constant and that the behaviour of one or two individuals under a particular set of conditions cannot be taken as indicative of the behaviour of the herd from whence they came, let alone animals generally.
>
> (Ivins, 1952)

Similarly, Arnold (1981) warned about preferences of animals in cafeteria-type tests. He indicated that a considerable amount of information may be gained about the preferences of sheep for the particular plants under study, but there is no real indication of differences between potential intake of plants or of preferences for these plants in mixed swards. These studies need to be continued over seasons and with various sward conditions.

Current studies of diet selection have built a theoretical framework from which many experiments could be done to examine further the relative importance of social models and post-ingestive consequences on diet selection. It would appear, however, that we are still a long way from any predictive model of diet selection and hence food intake.

Further Reading

Arnold, G.W. and Hill, J.L. (1972) Chemical factors affecting selection of food plants by ruminants. In: Harborne, J.B. (ed.), *Phytochemical Ecology*. Academic Press, London, pp. 71–101.

Baile, C.A. and McLaughlin, C.L. (1987) Mechanisms controlling feed intake in ruminants: a review. *Journal of Animal Science* 64, 915–22.

Chapple, R.S. and Lynch, J.J. (1986) Behavioural factors modifying acceptance of supplementary foods by sheep. *Research and Development in Agriculture* 3, 113–20.

Hodgson, J. (1982) Influence of sward characteristics on diet selection and herbage intake by the grazing animal. In: Hacker, J.B. (ed.), *Nutritional Limits to Animal Production from Pastures*. Commonwealth Agricultural Bureau, Farnham Royal, pp. 153–66.

Penning, P.D., Parsons, A.J., Orr, R.J. and Treather, T.T. (1991) Intake and behaviour responses by sheep to changes in sward characteristics under continuous stocking. *Grass and Forage Science* 46, 15–28.

Provenza, F.D. and Balph, D.F. (1990) Applicability of five diet-selection models to various foraging challenges ruminants encounter. In: Hughes, R.N. (ed.), *Behavioural Mechanisms of Food Selection*. NATO ASI Series G, Ecological Sciences, Vol. 20. Springer Verlag, Heidelburg, pp. 423–59.

Provenza, F.D., Pfister, J.A. and Cheney, C.D. (1992) Mechanisms of learning in diet selection with reference to phytotoxicosis in herbivores. *Journal of Range Management* 45, 36–45.

2

Social Behaviour and Organization

Introduction

The dominant social characteristic of sheep which is recognized by anyone who is familiar with the species is their 'flocking nature'. It seems likely that this is one of the characteristics which allowed man to control groups of sheep and begin their domestication. Kilgour (1976) has described the sheep as:

> a defenceless, vigilant, tight flocking, visual, wool covered ruminant; evolved within a mountain grassland habitat, displaying a follower type dam–offspring relationship with imitation between young and old in establishing range, showing a seasonal breeding and a separate adult male sub group structure.

Add to these characteristics the relative ease of taming and it would seem that this animal is ideally suited to control by man, at least in a grassland environment.

There is a wide diversity of environments in which sheep are found and, at first glance, a wide variety of social structures are apparent for different species. Many different social structures appear to exist between breeds and species as well as within breeds and this variety may well relate to their origins. Some authors suggest that all domestic sheep originate from the Asiatic Mouflon (*Ovis orientalis*) which implies that behavioural differences observed between domesticated breeds may be due to selection by man for different behavioural characteristics most suited to the environment in which the sheep were found. An example of this is the greater dispersal of hill compared with downs breeds in the UK. However, some of the sheep breeds presently found in the UK may have originated from the

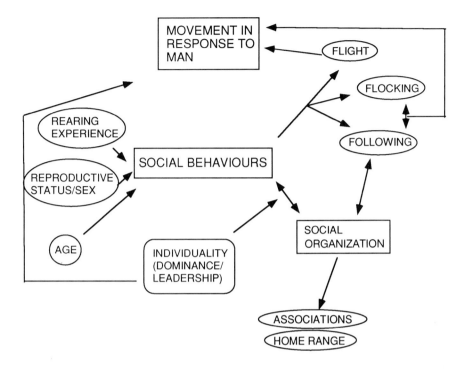

Fig. 2.1. The factors influencing sheep social behaviours and the links with social organization and movement in response to man.

Argali (*Ovis ammon*) and therefore behavioural differences may also relate to species differences.

Whatever the origins of the different breeds, as we begin to compare behaviours it soon becomes apparent that we are in fact dealing with a continuum of social organization with a surprising consistency in the basic behavioural patterns.

In this chapter we describe the major social behaviours of the sheep, its social organization and the mechanisms which contribute to its maintenance. The knowledge of these basic behavioural characteristics will then be related to handling sheep as individuals and as a flock; Fig. 2.1 illustrates the links between behavioural characteristics of the sheep and the sheep's response to man.

A number of reviews have linked the social organization of ungulate species with their feeding strategies. Having said this it is also clear that social organization of species must remain flexible enough to allow adaptation to changing seasonal conditions or changes in resource distribution. This is well illustrated in the behaviour of Bighorn sheep in a desert environment where the normally seasonal breeding animals have adapted their social patterns to the unpredictable nature of the desert by the males remaining with the females throughout the year (Lenarz, 1979).

In seasonally breeding populations such as the sheep, social organization often undergoes cyclic changes with organization usually being most complex during the rut. This may be simplified by man's modification of resource availability (transhumance systems of husbandry).

The means by which social organization is established and maintained is primarily through a system of species-specific communications, although there are many common visual displays used by ungulate species. Our understanding of these communication systems is sketchy but this chapter also gives some insights into the communication systems used by sheep.

Mechanisms Maintaining Social Structure

Individual or group recognition

It is difficult to prove the existence of recognition in large groups of animals but it can be argued that the maintenance of group cohesion would be impossible without such recognition. There is clear evidence of individual recognition between mother and young. In this case it appears that the signals for recognition involve olfactory, auditory and to a lesser degree visual and tactile stimuli. Whether the same mechanisms of recognition remain as the animals grow older and how many individuals an animal can identify are questions which remain to be answered.

The fact that flocks of ewes reared separately or flocks of different breeds remain segregated for many months after they are mixed suggests that there is also a mechanism for group recognition within sheep flocks.

Mechanisms for recognition

Leuthold (1977), in his book on African ungulates, outlines a classification of animal signals based on their function. This may include: attraction (courtship), repulsion (threat), submission and disturbance (alarm). The initial three functions all provide information about social status. The various forms of signals which can transmit this information are visual (posture or movement), auditory (vocalizations), olfactory (pheromones) and tactile (contact). In general, the complexity of these signals increases with the complexity of the social organization.

Visual signals

Facial expressions which are commonly recognized as of great importance in primates, are not often used by ungulates who mostly rely on postures

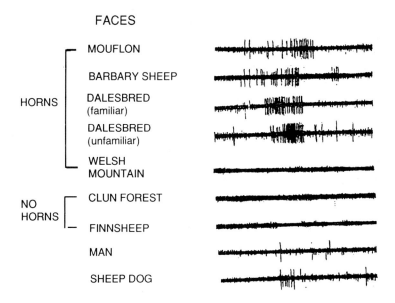

FACES

MOUFLON

BARBARY SHEEP

HORNS

DALESBRED
(familiar)

DALESBRED
(unfamiliar)

WELSH
MOUNTAIN

NO
HORNS

CLUN FOREST

FINNSHEEP

MAN

SHEEP DOG

Fig. 2.2. The electrical response of sheep brain cells to head profile stimuli
including horned and non-horned animals (data from Kendrick and Baldwin, 1987).

and gestures (pantomimics). Static postures include lateral or broadside
displays, or simply, the movement or positioning of the head. Dynamic
signals may include horning of the ground, pawing, and stamping.

Sheep are predominantly visual animals with alarm initiated by
individual animals who make visual contact with a 'predator'. In feral
populations there is evidence of the 'alert' being initiated by vigilant
individuals. Alert and attention postures are also seen in domestic sheep
but normally only in relatively homogeneous environments lacking shelter
or in flocks of mixed age and sex.

Although many of the visual signals recognized by sheep rely on postures or dynamic signals there is some evidence that sheep can recognize
shapes and that various regions of the brain respond differentially to
various social stimuli. A study by Kendrick and Baldwin (1987) showed
that in Dalesbred sheep, groups of cells in the temporal cortex responded
differentially to images which included the faces of: unfamiliar sheep of the
same breed, horned or non-horned sheep, a dog and a man (Fig. 2.2).
These data demonstrate that sheep do have the ability to distinguish
between objects which may have social significance. One group of brain
cells was sensitive to horn size, another to familiar animals (same breed)
and a third to threatening stimuli (human or dog).

Such observations present the possibility that visual identification of

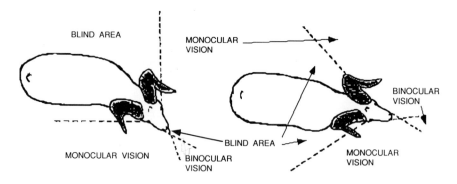

Fig. 2.3. The visual field of the sheep.

other animals as a group may play a major role in social organization. Whether recognition is at an individual animal level is unclear although recent studies have shown that sheep can identify people responsible for causing pain for a period of months after the event (Fell and Shutt, 1989). The possibility of individual recognition is also supported by the use of vision in mother–young identification. Shillito-Walser (1978) showed that a number of breeds used sight as a major mechanism for location and recognition of lambs by ewes. Sight was apparently more important than auditory cues. However, sheep quickly changed to using sound if sight was restricted.

Sheep have a wide field of vision with a blind area of only 90° behind the head of a moving sheep (Fig. 2.3). There is some evidence for colour vision, with ewes discriminating between lambs of different colours particularly colours in the longer (dark) wavelength. Sheep also appear to have an extremely good depth of perception.

Auditory signals

Sounds include those made by the vocal chords as well as snorts, whistles and grunts. The repertoire of vocal sounds used by sheep ranges from a 'rumbling' sound made by ewes toward newborn lambs and also rams during courting, the 'snort' of aggression or warning to the 'bleating' of contact and distress calls. Sheep have a wide range in auditory sensitivity, particularly in the upper frequency ranges.

Vocal communication may be used in a number of different settings including mother–young contact, mating, territorial warning, alarm, aggression and distress. Auditory communication is predominantly associated with mother–young bonding. However there is incidental evidence of communication via sounds in older animals. The level of vocalizations

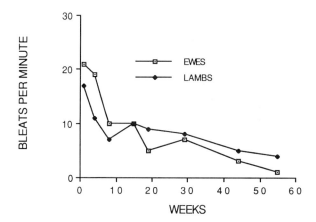

Fig. 2.4. Vocalization levels recorded during recognition tests of ewes by their lambs at various ages (data from Hinch *et al.*, 1990).

exhibited by Merino dams and their offspring declines to a very low level 10 weeks after parturition (Fig. 2.4). This low frequency of vocalizations is also apparent in yearlings. Similar declines in vocalization with age of animals have been reported in the Soay.

Studies by Shillito-Walser (1978) suggested that auditory cues are used for orientation rather than specific identification of lambs in Clun Forest, Finnish, Jacob, Dalesbred and Soay breeds. Subsequently, Shillito-Walser and Hague (1980) showed clear differences in the sonographic characteristics of Clun Forest, Dalesbred, Jacob and Border Leicester breeds; Clun animals having the most heterogeneous bleats and Jacobs the most homogeneous. It was also suggested that frequency of bleating may be the means whereby ewes identify their own from alien lambs.

There is evidence that breed recognition of lambs by ewes is to some degree influenced by auditory signals. Ewes were more responsive, in terms of rapidity of approach and number of bleats, to recorded lamb calls of their own breed rather than alien lambs. This recognition of breed was present whether the lambs were 15 or 80 days old (Shillito-Walser *et al.*, 1982).

The frequency of bleating of the Soay sheep is much lower than that of Dalesbred or Border Leicester ewes with Soay tending not to respond to bleats in a regular manner. Dalesbred and Jacob sheep use vocalizations in different ways; the Dalesbred utilizing vocalization more specifically for communication with their own lambs, and being more accurate in identification of their lambs.

For the lamb the 'meaning' of ewe bleats may change with age, initially serving as a 'call' and later simply as a means of location. The observation that the lambs of the domesticated Dalesbred respond to ewe calls for at least three months led to the conclusion that the incidence of vocalization may have been altered by domestication with a higher incidence of vocal communication used in the domesticated breeds compared with the feral types such as the Soay. Comparison of domestic breeds is necessary to confirm this suggestion. It is true that there is considerable variation in the frequency of bleating between breeds of Cheviot, Romney and Merino origins.

Vocalizations may be used in adult animals as a sign of distress. In Soay sheep on St Kilda Island Grubb (1974a) recorded vocalizations in situations of distress; 'when one animal had lost its companions. It would look around, bleat and move hesitatingly and erratically for a short period'. The increase in vocalizations during distress is a characteristic which has been used by some workers in the assessment of individual sheep's responses to isolation.

Olfactory signals

Most of the evidence for the role of olfaction in the maintenance of social structures in ungulates is circumstantial but it is clearly an important means of communication. A particular role is the transfer of information to conspecifics about territory, individual and group recognition and the establishment of sexual contacts.

In sheep the role of olfaction in the recognition of groups or flocks has been suggested by Arnold (1985) who showed that sheep rendered anosmic had no capabilities for group recognition. Sheep can discriminate between wool, faeces, saliva and infraorbital secretions of different origins, but how important such discrimination is to the maintenance of social structure is unknown. In recent studies of Merino ewes there was no evidence that shearing reduced recognition of nearest neighbours, or subgroups (Fig. 2.5).

Individual animal recognition clearly involves olfaction in mother–lamb recognition and it is likely that it may also be important in the maintenance of short-term associations within sheep flocks although this has not been proven. A major role of olfaction in sheep is in the recognition of the sexual state of individuals which will be discussed in greater detail in Chapter 3.

In some ungulate species, scents of urine and faeces are used to mark out territory, but there is no evidence for this behaviour in the *Ovis* species although some scent glands are present in the sheep. No response of ewes has been noted where interdigital scents have been deposited on grass or where interdigital secretions have been tested as an attractant. Antorbital or infraorbital glands are used for deliberate marking by some ungulate species, particularly territorial species. Evidence of marking in sheep is scarce although marking of fence posts by matriarchal ewes has been

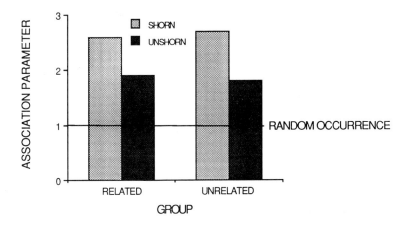

Fig. 2.5. Recognition (as measured by frequency of association) of related and unrelated sheep (reared together as flock-mates) before and after shearing.

recorded near to feeding areas (Stolba *et al.*, 1990). Such behaviour was to some extent dependent on the presence of a heterogeneous environment although the most complex social group (family groupings) exhibited marking behaviour in all environments.

Data from wild ungulates suggest that scent marking is most commonly used in animals of 'high' social status and is also often associated with agonistic encounters, presumably conveying information about territory or status. The observations of Stolba *et al.* (1990) would support such a hypothesis for sheep, as almost all marking behaviour was within an area around the 'feeding site' of the paddock.

Tactile signals

Physical contact between animals can be a means of transmitting information about social status. Grooming is recognized in cattle to be a form of behaviour which can be linked to social status, with lower-ranked animals more likely to groom higher-ranked animals. This form of communication appears to be of minor importance in sheep. However, in the final determination of social status, physical contact becomes important with butting and head clashes common among rams, particularly before the rut.

Social Behaviours

Having established the communication mechanisms of the sheep we will now look in some detail at the social behaviours exhibited by sheep and how these behaviours are used in the organization of social structure and spatial distribution.

Agonistic behaviours

This classification of behaviour includes all forms of aggressive behaviours as well as non-aggressive responses such as defence, appeasement and submission. The more vigorous forms of aggressive behaviour are restricted in their incidence in most species of ungulates including sheep with display and intimidation rather than fighting being the more common forms of agonistic encounters. Associated with these intimidatory displays are usually the appeasement or avoidance responses which allow major conflicts to be averted.

Geist (1966) has suggested that horns have evolved as an organ of social conflict and argues forcefully from the example of the Mountain sheep that it is horn size that is a major determinant of social status in a sheep flock, status being maintained via horn display. This hypothesis has physiological support in the findings of Kendrick and Baldwin (1987) who reported specific brain cells responsive to the presence of horns.

Threat can involve a range of signalling behaviours, such as, lateral or broadside display or frontal display using head and neck size (Leuthold, 1977). In polled sheep breeds or where females are hornless it is likely that these signals become important in the maintenance of social status.

Submissive behaviours can vary from the extreme of fleeing to the signalling of acceptance of defeat or unwillingness to engage in conflict by a lowering of head and neck. Such behaviour allows subordinate animals to stay in the vicinity of the dominant while avoiding conflict and possible injury. Squatting and urination behaviour has been described in female mountain sheep as a submissive posture in response to butting or threat and it is possible that scent marking may not always be linked with such behaviours.

The main agonistic behaviours of *Ovis* species have been summarized by Schaller and Mirza (1974) in their comparison of the behaviours of the Punjab Urial (*Ovis orientalis punjabiensis*) with other wild sheep such as Mountain sheep, Soay, Marco Polo and Mouflon. These behaviours have also been fully described by Geist (1971) for the Mountain sheep. Some 20 agonistic behavioural patterns have been recognized in sheep populations with the majority exhibited by rams. These behaviours are most often exhibited by animals of similar age or horn size. Schaller and Mirza (1974) report that 90% of interactions were between rams of similar age in their study of Urial rams. There is some variation between sheep species in the frequency of these behaviours, but whether these differences are definitive in being able to distinguish between species seems unlikely.

The main agonistic behaviours are summarized in Table 2.1 and are in

Table 2.1. The most common agonistic behaviours of wild and domestic *Ovis* species.

Type of behaviour	Urial	Mountain	Soay	Marco	Mouflon	Domestic
Kick	+	+	+	#	+	+
Head clash	+	+	+	+	+	+
Twist	+	+	+	+	+	+
Horn threat (jerk)	+	+	+	+	+	+
Nudge (poke)	+	+	?	?	?	+
Rub (horns or face)	+	+	+	?	+	+
Huddle	+	+	?	?	+	#
Butt	+	+	+	+	+	+
Jump (threat)	+	+	?	+	+	?
Horn pull	#	#	?	?	+	?
Shoulder push	#	+	+	?	+	+
Mount	+	+	+	?	?	+
Low stretch	+	+	+	+	+	+
Block	+	+	+	?	?	+
Head shake	?	+	?	+	+	?
Paw	?	+	?	+	+	+
Neck low	+	?	+	?	+	+

+ = pattern present
\# = pattern present but rare
? = pattern not observed.
Source: Schaller and Mirza (1974).

order of frequency of occurrence. Clearly there are some differences between *Ovis* species but these seem to be minor.

A brief description of the main behaviours divided into aggressive, threat and submissive characters follows.

Aggressive

Head clash is the best known behaviour because of its apparent ferocity as rams separate by 3–10 m before charging at one another clashing head or horns together (Fig. 2.6). After impact animals may stand facing one another until one submits and moves away or they charge again. A low intensity form of this behaviour can be found with rams close together (<1 m) jerking their head downward and toward one another until horns touch. Low intensity head butts have also been reported between ewes and are observed to be very common between Mouflon ewes (McClelland, 1991). Mountain sheep exhibit a jump prior to the clash which is not seen in the Mouflon or Soay or domestic breeds.

Fig. 2.6. Merino rams in charging position prior to a head clash.

Fig. 2.7. Scottish Blackface ewes exhibiting butting behaviour.

Butt and nudge involves a butt with the head into the side or rump of another animal or a marked poke with the muzzle (Fig. 2.7). The intensity seems to vary greatly.

Horn pull and shoulder push are behaviours which could be described as jostling with no apparent social outcome.

Blocking is an inclusive term used to describe a complex sequence of behaviours where animals in reverse parallel position or parallel head to head position engage in a prolonged series of behaviours including nudging, horning, shoulder pushing and butting. It usually occurs as part of an agonistic encounter between evenly matched rams.

Mount is included as it would appear to the be prerogative of dominant animals when exhibited within ram groups. It has been observed in ewes and wethers usually as part of a playful agonistic encounter. In such situations contact is usually brief. Mounting does not appear to be a visual signal for sexual activity as is the case in cattle.

Threat

Kick is an upward movement of a stiffened foreleg which can be repeated a number of times and sometimes contacts the opponent to which it is directed.

Horn threat (jerk) behaviour is expressed by rams normally in competitive situations and involves the movement of the head sharply downwards and towards an opponent.

Twist and low stretch involves the neck being held horizontal to the ground with the muzzle forward and raised while the twist involves the stretch followed by a turning of the head through 90° often with a flicking of the tongue. Both behaviours are thought to be forms of 'broadside display' and are commonly seen both in the context of male–male and male–female interactions. There seem to be differences in the incidence of these behaviours between *Ovis* species. They are rare for example in the Soay and Urial but common in the Mountain sheep and Mouflon.

Head up is exhibited by rams when they raise their head in such a manner that the neck bulges markedly and this, associated with an erect stance and stiff walk, further accentuates size.

Huddle behaviour is infrequent and is initiated by some form of aggressive interaction between one pair of animals. Rams congregate together with heads lowered toward one another and engage in a series of aggressive and threat behaviours. Sheep normally are not observed facing one another directly, but prefer to orientate themselves away. Presumably the head to head orientation is a threatening act and the normal outcome will be an agonistic encounter.

Submissive

One of the few basic behavioural differences which has been identified between the wild sheep populations appears to be the submissive postures of the various species. In the mountain Bighorn sheep of North America (*Ovis canadensis*) submissive postures seem to be absent, whereas they are present in animals of Urial or Mouflon origins. The Corsican Mouflon (*Ovis ammon*) appears to have a unique submissive posture in which the dominant animal kneels and is licked on the neck by the less dominant male.

Low neck posture appears to be a common form of submission and is seen in the context of retreat from an aggressive encounter between rams.

Head shake behaviour is exhibited almost exclusively by small animals in response to the presence of a larger animal. Normally the animal turns away from the larger before moving its head.

Amicable grooming

Rubbing of horns, face, muzzle or neck on another animal is relatively common particularly between rams within mixed sex groups. It appears to be tolerated by most animals but definitive data for single sex or age groups is lacking. The behaviour has been recorded in single age groups but at a low frequency. It seems likely that this behaviour spreads pre-orbital secretions from one animal to another.

The nibbling or licking of rubbed areas is not an uncommon prelude to rubbing behaviour. Grooming also appears to be more frequent in young lambs and yearlings as well as in ewe and neonate. In 30 hours of detailed observations of a Soay ewe and her six-month-old lamb, the lamb was observed to spend 9% of its time grooming other animals while grooming time of the ewe was negligible (Grubb, 1974a).

Differences between sexes

The social repertoire of the female is clearly smaller than that of the male with only about 50% of the behaviours of the male Bighorn being exhibited by females (Eccles and Shackleton, 1986). This may be due, in part, to the lack of variation in horn size in the females with the consequent lack of horn displays but also to the markedly lower incidence of aggressive behaviour. In the Bighorn the only horn-orientated activity exhibited by females appears to be hornpull, a behaviour occurring predominantly as a response to a butt.

The reason for such a markedly lower incidence and shorter duration of agonistic encounters in ewes is difficult to determine but the pattern is widespread within the *Ovis* species (Schaller and Mirza, 1974). It may reflect the fact that food resources and sexual partners are rarely limiting and consequently, competition for resources is not necessary. In the more gregarious domestic breeds agonistic encounters between females are also rare and it is only in situations of individual competition for feed that aggression is clearly exhibited. However in a study of a small group of two-tooth Scottish Blackface ewes, relatively high levels of aggression have been reported, particularly when the animals were resting in shelter (Lynch *et al.*, 1985). Reasons for this may be related to the age of the animals, with younger animals still establishing stable social structures, or it may reflect competition for the relatively limited shelter resource.

Geist (1971) has suggested that in terms of social behaviour, adult rams treat all other members of a flock as subordinates and do not distinguish between sexes, except on the basis of size. This concept rests on the fact that similar social behaviour is exhibited by females and juvenile animals within a flock. This appears to be true for anoestrous females with oestrous females behaving more like subordinate males.

Geist's concept is supported by data from a domestic breed where Merino flocks of different age, sex and family composition showed aggression levels which were markedly lower in a 'family' group (ewes and two generations of offspring) than in the mixed age–sex group (ram, wethers and ewes), (Stolba *et al.*, 1990).

Differences between ages

There is evidence that in wild and domesticated sheep breeds the frequency of aggressive behaviours increases with age. In wild sheep, Schaller and

Table 2.2. The frequency (%) with which males of different age classes exhibited some agonistic behaviours.

Behaviour	Age (years)				
	< 1	1–2	2–3	4–7	> 8
Head clash	1	13	16	18	52
Kick	0	2	7	20	71
Twist	0	6	10	20	64
Low stretch	0	0	0	33	67
Butt	8	17	0	17	58
Rub	5	19	28	19	29

Source: Schaller and Mirza (1974).

Mirza (1974) showed that 52% of head clashes occurred between mature males with only 1% being exhibited by yearling males. In contrast, rubbing was common in all age groups (Table 2.2). The aggression levels between adult mountain rams tend to be lower than in younger animals while the level of threat/display is relatively much higher.

Attention and alarm

Attention behaviour has been described in feral, wild and domestic sheep. It consists of a 'frozen' posture (alert) with the animal staring in the direction of the disturbance (Fig. 2.8). The alarm behaviour consists of the animal moving with a very rigid gait usually with head raised. The incidence of attention behaviour was observed to vary according to the environment in which Merinos were located (Stolba *et al.*, 1990) with few topographical features such as shelter resulting in a greater level of attention behaviour (Fig. 2.9). In mixed-age or mixed-sex groups, older animals dominated the incidence of attention behaviour. This pattern was not true in a group based on 'families' which exhibited a low incidence of such behaviours.

Reports of wild and feral sheep leave the impression of a highly vigilant species which, when frightened, will normally flee. The initial movement of a few individuals results in the movement of the whole flock.

Fig. 2.8. Attention or alert position of a Cheviot ewe.

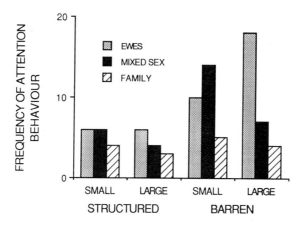

Fig. 2.9. The mean frequency of attention behaviours exhibited by three groups of sheep in small (0.24 ha) and large (1.2 ha) fields with (structured) or without (barren) trees and slope (data from Stolba *et al.*, 1990).

Exploratory and following behaviours

Exploratory behaviour is common in sheep introduced to a new area. However, the distance of movement may initially be restricted to a small area around the entrance; the rate of dispersal on entering an area varies with breed.

The 'following' nature of the sheep often results in them walking in single file as animals move from one area to another or to water. The formation of tracks is commonly observed radiating from camp sites or watering areas. A number of observers suggest that paths may follow a slight weaving pattern rather than a straight line. This may be the result of the lead animal attempting to retain visual contact with the animals behind. Alternatively sheep may maintain visual contact by moving in several parallel lines.

This 'following' behaviour is commonly observed in sheep flocks throughout the world. Grubb and Jewell (1966) describe the activity in Soay:

> sheep would keep in single file and travel along well defined
> tracks. If the leader of the file halted, so would the others. The file
> would periodically break up as pockets of good grazing were
> encountered. Files were never led by very young sheep if older
> ones were present; activity was generally initiated by older animals.
> Though no individuals stood out as 'flock leaders', some older
> sheep were clearly more alert and wary than others.

Breed differences are apparent in the degree of expression of 'following' behaviour but all breeds establish following behaviour between mother and young. This pattern is retained in adult life between conspecifics.

Play behaviour

Adult Soay sheep have been reported to exhibit apparently pointless behaviours such as trotting, galloping, head tossing, rearing and bucking during the spring. Grubb (1974a) suggested that this was associated with a recovery phase after winter 'starvation'. Play behaviour has also been observed by the authors in both young ewe groups and in mature non-pregnant ewes of Romney, Border Leicester and Merino breeds. It seems to be most prevalent in groups of adult animals in positive energy balance and is also commonly observed in lambs a few days old to ten weeks of age.

The majority of play behaviours are exhibited in a social context. It is possible that, functionally, play, both in lambs and adults, involves the establishment and development of the 'motor' actions necessary for social and sexual interaction and animal maintenance.

Sheep have a rich repertoire of social behaviours, particularly behaviours with visual impact. It is apparent that males express most of these behaviours far more frequently than females. However, all animals and particularly older ones, exhibit a vigilance to disturbance and a willingness to respond to the flight of another flock member. Sheep also show 'follower' behaviour both between mother and young and as adults.

Social Attributes of Individuals

Having outlined the major social behaviours exhibited by sheep, the next question to be addressed is how these behaviours are used by the animals in some form of social organization. It has been suggested that organization could be based on attributes of individuals including dominance, leadership and associations between animals. These attributes are now examined in a more general context to evaluate their role in the organization of sheep flocks, be they wild, feral or domestic breeds.

Dominance

This concept has been defined in various ways but is in essence 'the attribute which provides the holder with access to a resource or resources in precedence over others and without contest' (Leuthold, 1977). There has been much discussion of this concept since it was first defined but it seems that dominance is probably quite specific to the control of one resource. This can include social priorities such as rank or territory, sexual priorities

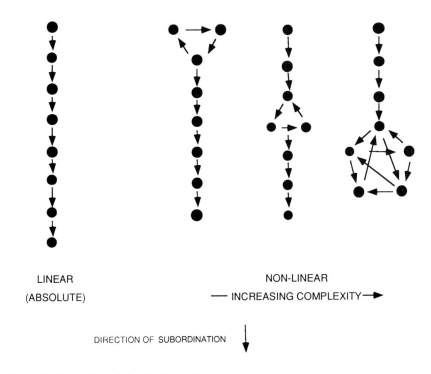

LINEAR   NON-LINEAR

(ABSOLUTE)   ── INCREASING COMPLEXITY ➤

DIRECTION OF SUBORDINATION ↓

Fig. 2.10. Examples of simple hierarchy structures.

such as mating rights, or priority access to food, water and shelter.

Dominance, established via threat and aggression, is formed rapidly in many ungulate species. In sheep, hierarchies are often bidirectional and non-linear so that dominance of one sheep over another is not absolute (Fig. 2.10). Overt aggression is rare between extremes in a dominance hierarchy with most contests occurring between animals of similar rank.

Measurement

Dominance is normally determined by recording the agonistic encounters between pairs of animals and counting the proportion of wins and losses between these pairs. A dominant animal is usually one which wins more than 50% of the time (rather than an absolute dominance). From this information a ranking or dominance order can be determined. Craig (1986) has evaluated a number of methods used to assess dominance but very few studies have used more than one method in appraising dominance status.

Dominance in sheep

In sheep, social rank is not as obvious as in some other species. This is particularly so in domesticated flocks where groups are often composed of single sex or age groups and where agonistic encounters are relatively rare. Little is known of the way dominance develops in domesticated sheep flocks but it appears to be determined, or at least influenced, by such factors as age, size and weight, sex and aggressiveness. Dominance almost certainly functions as a means of reducing overt aggression and injury in flocks.

Dominance orders and hierarchies have been identified in sheep flocks for a number of competitive situations particularly with rams and wethers. Recent studies (Lynch *et al.*, 1989) of Merino ewes suggest that dominance is not a clear-cut phenomena in this breed nor in Merino crosses. In fact it

Table 2.3. Dominance ranking in a 'family group' and a mixed-sex group of Merino sheep (in the family group the symbols ' and " represent 2.5- and 1.5-year-old offspring of equivalent numbered ewe; in the mixed-sex group the first digit represents age of the animal – all males except the ram were wethers).

	Family group			Mixed-sex group		
Rank	Animal number	Sex	Weight (kg)	Animal number	Sex	Weight (kg)
1	1'	M	51	Ram	M	68
2	1	F	47	30	M	71
3	15	F	48	50	M	69
4	11	F	59	32	F	58
5	11'	M	48	31	F	55
6	9	F	53	51	F	60
7	15'	M	41	33	F	64
8	3"	F	39	10	M	39
9	7	F	53	13	F	42
10	11"	M	54	12	F	49
11	1'	F	38	11	F	36
12	15"	F	45			
13	9"	F	47			
14	7"	F	42			
15	3	F	51			
16	3'	F	39			
17	7'	F	39			
18	9'	F	36			

Source: Stolba *et al.* (1990).

has been argued that dominance is not important in the organization of single-aged ewe flocks. Similarly Grubb (1974a) describes the feral Soay in the following terms: 'A social hierarchy could not be observed amongst Soay ewes but they have nothing to compete for'.

In a flock of Merinos which included ewes, wethers and a ram, dominance, as determined by access to a limited feed resource, was influenced by sex and size with the ram being dominant over wethers and older females over yearling males and females (Table 2.3). Geist (1971) has suggested that dominance rank in Mountain sheep is determined by horn size and is independent of sex but the former study (Table 2.3) suggests that size and weight may be important in some sheep groups where animals are polled.

The implications of social status for animal productivity are not clear for sheep but evidence from operant conditioning studies indicates that animals of low status have greater difficulty coping with new environments, human handling and space restrictions than animals high in the dominance order.

Leadership

This concept is often confused with that of dominance but they are two distinct entities. Leadership is expressed by animals which initiate movement and an association between leadership and dominance is difficult to establish in most ungulate species. In most studies of single sex or single age groups of sheep the correlations seem to be slight or non-existent. This may be due to the diverse functions which the two attributes appear to have. Leadership is a behaviour which functions to maintain knowledge of an environment, perhaps via mother–young associations, and consequently functions to co-ordinate group cohesion in terms of movement to food and water. Such a function leads to the follower pattern already described.

Examination of male and female flocks of sheep have shown that there is evidence of individual animals initiating movements, but the animal which is the leader differs according to the reason for the movement. For example, a different sheep will lead a forced movement compared with a voluntary movement. Most studies of voluntary leadership have found no consistent order but Squires and Daws (1975) found some consistency of voluntary movement order for Border Leicester and Merino wethers when they moved along a laneway to water. There was an element of dominance in this experiment in that access to water was restricted. In general males are more likely to have a consistency in movement order than females.

In a detailed study of young Merino ewes only the first and last animals were consistent in either voluntary or forced movement orders and there was no evidence of a relationship between these rankings and dominance. Similar results have also been reported for castrate males of various breeds.

In sheep flocks, leadership is very much an individualistic behaviour reflecting a lower gregarious drive of the individual animal taking the lead. The possibility that such individuality may be also expressed in other behaviours was evaluated by Lynch *et al.* (1989) who examined the relationship between fear, exploratory behaviour and leadership in young Merino and Merino × Border Leicester ewes. Repeatability of these tests was poor and no relationship with leadership order was apparent. It was concluded that individuality was not consistent across behaviours and that leadership was not a robust phenomenon particularly in forced movement situations.

It has been argued that forced movement order may be related to the ability of animals to tolerate social disruption caused by yarding, rather than to the establishment of a priority order. Syme (1981) tested this hypo-thesis by selecting sociable and non-sociable animals based on their level of activity during isolation, and mixing them together as a flock before testing for movement order and leadership. Consistent positions were established particularly for the non-sociable group which were normally to the rear of the movement order. Leadership appeared to be shared between two or three of the 'sociable' animals.

There is some evidence to suggest that older animals may initiate move-ment within a flock. Davis (1938) in describing Bighorn activities reported 'individualism was shown by some sheep, particularly by "Broken Horn" the old lambless ewe that led the flock on most occasions'. Similarly in the recent study of Merinos where mixed age groups were established, move-

Table 2.4. The proportion (%) of initiated movements related to age in (a) Merinos and (b) Bighorns.

(a) Merino	Age (years)		
	Oldest (4+)	Intermediate (2–4)	Youngest (< 2)
Flock A	44.3	23.8	29.0
Flock B	8.9	13.1	7.1
Flock C	44.0	–	31.0

(b) Bighorn	Age (years)			
	> 7	5–7	3–5	1–2
% showing leadership	97.5	25.0	8.5	6.0
Average group size led	6.1	5.4	2.9	2.5

Source: Adapted from Stolba *et al.* (1990) and Geist (1971).

ments were initiated more often by the older members of groups but leadership was not exclusive to one animal in any of the four flocks studied (Table 2.4a).

Responses to leadership are also influenced by age and by environmental conditions. One report showed that older animals followed initiators of movement more often than did younger animals. The same study showed that a more heterogeneous environment induced a greater number of animals to follow, possibly because of an increased responsiveness to other animals in a situation where vision was restricted.

Grubb (1974a) reported that leadership in Soay sheep was not conspicuous although earlier observations suggested that movement was predominantly led by older animals. In most reports where leadership is given by older animals it is also noted that animals tend to 'defend' their position by butting animals who may try to pass.

In the Bighorn, leadership in males is associated with horn size (Table 2.4b) but this pattern has not been confirmed for other wild or feral populations of sheep. Again it is possible that leadership is a reflection of age and individuality of the older males rather than being directly related to dominance.

Associations

Associations between pairs of animals have been reported for many ungulate species. There is a hierarchy in these social attachments or associations, with the strongest being between ewe and lamb, then between twins, filial groups and finally peers.

In most studies, associations have been identified by the distance to nearest neighbours (NN), and in approximately 60% of cases the nearest neighbour is reciprocal for both animals. Arnold (1985) has suggested that these bonds may be stronger than group cohesion and consequently allow sub-group formation as environmental conditions dictate. Nearest neighbour distance varies considerably according to the animals involved but is normally < 5 m. The distance to the second nearest neighbour is far less predictable and is possibly more often related to flock dispersal than to associations with other animals.

In sheep there is mounting evidence of pair bonds between unrelated animals in a diversity of domestic breeds (Arnold, 1985). However, these bonds appear to have a relatively short lifetime of one or two months. In one study it was apparent that pair bonds as measured by NN distances, differed for the different activities of resting and grazing. Other studies show that pairing in twins is common, but occurs in only about 50% of potential pairs. Clear-cut associations are more common in mixed flocks than in single sex or age groups. There are also breed differences with Dorset and Merino animals forming relatively strong associations which are not observed in Southdown animals.

Evidence that pair associations result from rearing experience rather than genotype was presented in a series of studies using a mix of diverse breed types (Shillito-Walser *et al.*, 1981). However, there is also evidence of breed differences in the strength of association between dams and their offspring. For example, Dalesbred lambs were observed to spend more time with their dams after weaning than did Jacob lambs.

The breakdown of lamb–ewe associations

There are marked breed differences in the rate at which the breakdown of lamb–ewe associations occurs. Comparison of Jacob, Dalesbred and Soay ewes shows a markedly greater tendency for Soay ewes to return to their peers rather than to remain with their lambs 40 days post-partum (Shillito-Walser *et al.*, 1983). Associations between Scottish Blackface ewes and lambs weaken to the point of being insignificant by one year of age (Lawrence, 1990).

Associations between Merino ewes and their offspring can last for at least 2.5 years and even after a daughter has lambed (Hinch *et al.*, 1990). Nearest neighbour data suggested that the intensity of this association diminished over time with the lamb showing increasing preference for peers.

In a study of families within a large Merino flock (290), evidence was found for the maintenance of family members as nearest neighbours and also of preferences for animals from the group of origin. This will be discussed further in the context of sub-group formation.

Social Grouping and Spatial Relationships

Groups of animals usually consist of conspecifics in the same location and are of two types. Open groups have different individuals joining and leaving the group which leads to instability in size, age and sex composition. Closed groups have a stable composition and size. Members of these groups retain some spatial relationship to one another with individuals of even the most gregarious of species distancing themselves in relationship to one another. Such behaviour has led to the development of a number of concepts which help to define spacing. These concepts are illustrated in Fig. 2.11.

The term *individual distance* refers to the minimum distance to which animals would approach one another. This concept was refined by McBride (1971) who suggested that this distance is greater around the head and therefore he used the term *personal field* to define an area around an animal rather than a distance between animals.

The term *social distance* is used to define the maximum distance of

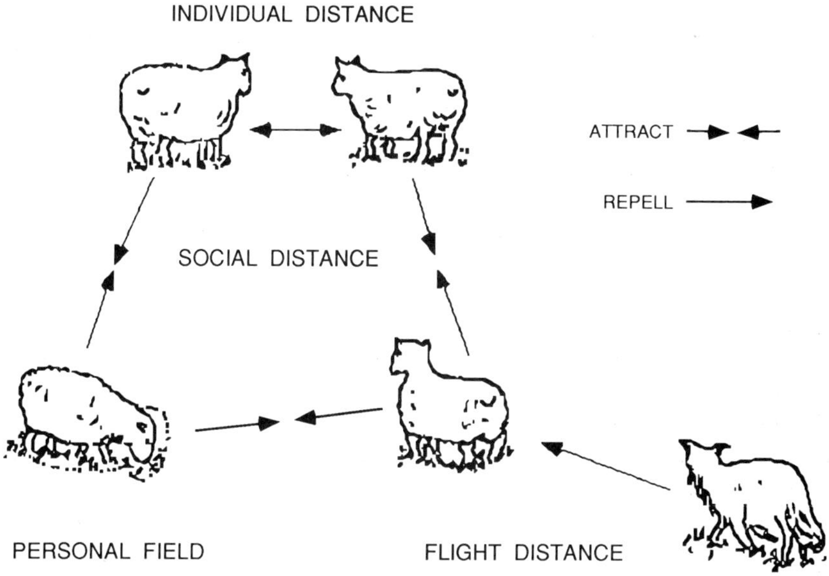

INDIVIDUAL DISTANCE

ATTRACT ➤ ◄

REPELL ➤

SOCIAL DISTANCE

PERSONAL FIELD   FLIGHT DISTANCE

Fig. 2.11. The spatial measurements used to describe social organization.

dispersal. It is a measure of cohesion between individuals of a group so that they maintain social contact.

Flight or *critical distance* is the third distance measure which can be used to describe spatial behaviours. This is the distance to which a 'predator' can approach before animals turn and flee and is of some importance in the context of handling.

The balance between individual distance and social distance in many ways determines the structure of social groups. Species which have short individual distances and have relatively small social distances are the species we define as *gregarious* animals. In most cases these species are found in large groups. In contrast, animals with large individual and social distances are often territorial animals who form small groups and defend a fixed area.

Dispersal

Ungulates show the following different forms of dispersal:

1. the solitary animal;
2. territorial organization where animals defend a particular area;
3. 'home-range' organization where animals limit their activities to a particular area;
4. animals which show no preference for areas, but distribute themselves in relation to one another rather than to a particular location.

Dispersal patterns are known to change in some species according to season and reproductive status. A number of authors have suggested that each of these patterns has developed to allow anti-predator strategies to be adopted in relation to the location of their primary food source or other ecological resources such as shelter.

The sheep's dispersal patterns fit broadly into the gregarious type species. If the broad categories of Jarman (1974) are used, the sheep is classified as a species of medium body size, with grassland habitat, and shows a considerable degree of selectivity in its diet. Consequently a social grouping would be predicted as having a flexible home range, being variable in size (5 to 200+) but with male adults forming male groups and females forming open groups, with gradual disassociation of mother and young.

This description of social organization is in fact close to that identified in wild and feral sheep studies. Wild sheep form combined female and juvenile groups located in home range areas. Some authors also suggest that the home-range group may consist of subgroups of related individuals ('families'). Males associate in bachelor flocks in another home range area separated from the females and juveniles. These 'herds' break up during the rut and males wander, joining a number of different female flocks during the 'rut'.

Flocking and group size (cohesion)

The mechanisms which influence the flocking tendency of sheep are unclear although there is clear evidence of variation within the *Ovis* species. For example Boyd *et al.*'s (1964) description of the Soay sheep: 'at the sight of the dog the Soay sheep did not bunch as do commercial breeds, but made off in all directions at high speed' contrasts with Winfield and Mullaney's (1973) report that both Merino and Wiltshire Horn ewes flocked together when disturbed by a dog while still retaining breed group identity.

In sheep it has been suggested that dispersal is to some extent controlled by individuals attempting to retain other sheep within their field of vision. Crofton (1958) found that most Corriedale ewes were orientated so that two other sheep subtended an angle of about 110° to their own head position (Fig. 2.12) and suggested that social spacing was controlled by the breadth of the visual field. This finding has not been evaluated for different breeds nor in environments where visual contact is restricted by vegetation.

The nature of flock structures also seems to vary according to breed with a wide but uniform dispersal in many British hill breeds, subgrouping in breeds such as the Dorset Horn and a tighter flocking structure for the Merino (Table 2.5).

Baskin (1975) reported that for Argali sheep two to three body lengths

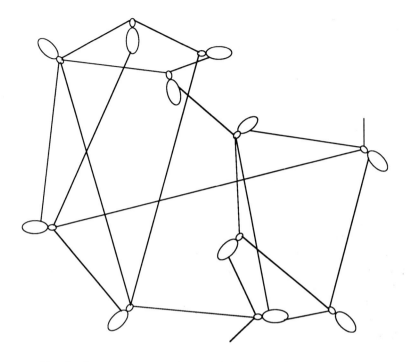

Fig. 2.12. The distribution of grazing sheep and the orientation of sheep such that two other sheep are within the 110° of the optic axes of the sheep (adapted from Crofton, 1958).

Table 2.5. Distances between nearest neighbours (NN) in sheep of different breeds.

Country	Sheep breed	NN distance (m)
Britain	Blackface	7.5
	Welsh Mountain	6.9
	North Country Cheviot	5.5
	Dalesbred	4.4
	Wiltshire Horn	3.4
	Suffolk	3.4
New Zealand	Romney	4.8
Australia	Merino	3.1
	Merino	1.5*

*Hinch (unpublished) for animals on good pasture.
Source: Arnold (1985).

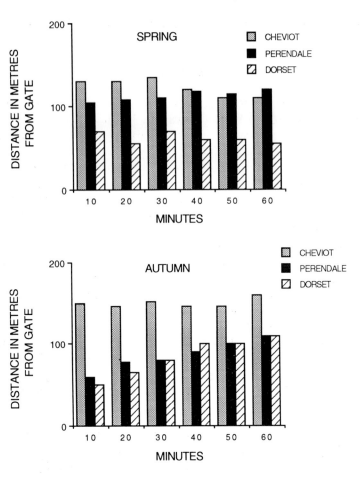

Fig. 2.13. The mean distance of dispersal from a gate of various ewe breeds (data from Kilgour *et al.*, 1975).

(1.7 m) was the approximate individual distance with a social distance of around 25 m. Studies of Merinos would suggest that individual distance is somewhat less (< 1.5 m) and that social distance is probably of the order of six or seven metres. The study of breed dispersal patterns reported by Kilgour *et al.* (1975) has shown that Perendale, Dorset–Romney and Cheviot breeds differ in their social distance as measured by dispersal from a fixed point. Season also appeared to influence dispersal but breed differences remained consistent (Fig. 2.13). This later observation may explain the difference between nearest neighbour (NN) distances of Merinos reported in Table 2.5. Arnold (1985) has suggested that the breed differences are in fact an adaptation to the environment in which the breeds developed. The mountain breeds of England seem to be at one extreme, moving about almost as individuals, while the Merino tends to respond only as a member of a flock.

Sub-groups

As well as differences in NN distances, breeds also differ in the cohesiveness of the total group. This is illustrated in Fig. 2.14 which shows a smaller number of groups with high dispersion or a larger number of groups with low dispersion, both covering the same total area.

As indicated earlier, minimal group size for sheep is normally four or five animals although this varies with breed and may not be evident unless the group is disturbed. In the case of parturient ewes the number may in fact be two (lamb + dam) while family grouping may range from two to six (Grubb, 1974a; Table 2.6).

The evidence presented earlier of associations between sheep leads to the further possibility that sub-groups of animals may form within a flock. These groups may consist of 'family' units or possibly peer groups. Peer groups of young ewes are rarely formed in Soay sheep, but peer groups of lambs and of young rams are relatively common even if transitory.

Arnold and Pahl (1967) recorded mean group size for adult and weaner sheep in two environments. Merino weaners tended to have smaller groups (mean of 33) compared with adult ewes (45). Breeds also differed, with Merino having the largest groups (33) and Dorset Horn × Merino weaners the smallest (10). This range is comparable with the 20 to 30 sub-group size reported for 'domestic' sheep in Russia (Baskin, 1974).

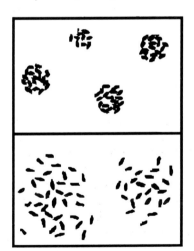

TIGHTLY GROUPED

SMALL NEAREST
NEIGHBOUR DISTANCE

WIDELY DISPERSED

LARGE NEAREST
NEIGHBOUR DISTANCE

Fig. 2.14. Sub-group formation in a flock of 60 ewes reflecting differences in cohesiveness and nearest neighbour distances (reproduced with permission from CSIRO, 1976).

Table 2.6. The frequency of observation of various sized family units in ewe home-range groups.

Year	Size of family units				
	1	2	3	4	+
1966	28	76	39	4	1
1967	42	54	21	10	4

Source: Adapted from Grubb and Jewell (1966).

Age influences group size with a gradual increase in group size with increasing age between four and 11 months. Under extensive grazing conditions, adult Merino ewes have been observed to form groups varying in size from six to 420 with a median around 100. Group size is reduced as pasture availability declines.

There is evidence that animals reared together may retain their group identity when mixed into a larger group. This is most clearly evident when different breeds are mixed. In the study of a mixed flock of Merinos and Wiltshire Horn ewes the identity of the two breed groups was retained for three months (Winfield and Mullaney, 1973) and similar patterns have also been noted for Dalesbred, Clun Forest and Jacob breeds. There is evidence of a gradual breakdown of breed links over time but breeds of similar flocking structure appear as the most likely associates.

The impact of rearing experience on the distribution of Merino ewes within a flock of 290 ewes was examined recently (Hinch and Lynch, unpublished). Four groups of ewes (18 to 59) were mixed with a group of two-tooth ewes and the distribution of focal animals from each group monitored three months after mixing. There was (i) a group which included three generations of six families, (ii) a group of four-year-old animals, (iii) a group of three-year-old ewes with some experience of group (i), and (iv) a group of two-year-old ewes taken from a larger flock.

For all groups, nearest neighbours were consistently members of their original group. The strongest associations were evident between family groups but interestingly, group (iii), which had been previously exposed to the family flock, preferred 'family' group animals as a 'second choice' (Fig. 2.15). Age groups showed a clear preference for animals of their own age and flock of origin. However, there was no evidence of utilization of particular areas by different groups. The animals grazed the paddock as one flock.

These data provide evidence for the fact that group identity can be

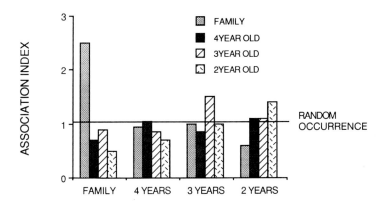

Fig. 2.15. The consistency of associations between different rearing groups of Merino ewes within a large flock.

recognized by sheep. The number of individuals which can be identified by any one animal is still unclear but it seems likely that group identity rather than individual recognition is the main cohesive force of sub-groups except in the situation where family groups are retained.

Home range

Jewell (1966) has described a home range as 'the area over which an animal normally travels in pursuit of its routine activities'. As such it is not the same as territory which connotates a defended area of particular social significance to the animal.

Home range size is suggested to be roughly proportional to body size (Leuthold, 1977) but food habitat, social structure and size of the social group may modify this. Migratory movement is not uncommon in ungulates, particularly in the more gregarious species, but the mechanisms which motivate these movements are unknown.

Geist (1971) suggests that there may be a number of seasonal home ranges in sheep, traditionally a summer and winter range. Males often have separate home ranges. Feral and wild animals appear to be loyal to their ranges and move in an orderly and predictable manner. Such movement is probably synchronized by seasonal changes in light in a similar manner to the influence of light on the onset of reproductive activity (Chapter 3). In domesticated sheep it seems likely that the movement between home ranges has been 'formalized' to some degree in transhumance patterns established by sheep herders throughout the world.

The existence of specific home range areas in sheep was first reported by Hunter and Davies (1963) for Scottish Blackface ewes which show a strong adherence to a particular area of the farm (Fig. 2.16). There are now a

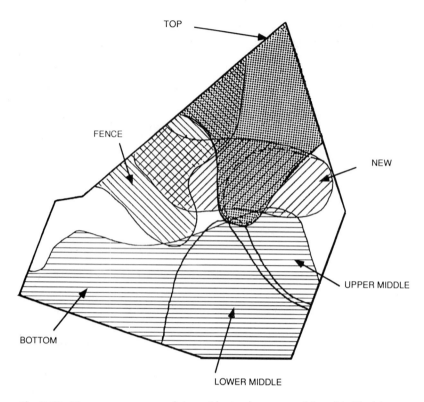

Fig. 2.16. Home-range areas of six resident sub-groups of Scottish Blackface ewes (adapted from Hunter and Davies, 1963).

number of reports of 'family' groups of two or three generations of ewes being identifiable within larger home range groups. One study identified groupings of two or three animals which were normally part of larger groups (17–160) in a feral population of Longwool-cross sheep on Campbell island. Another identified the family unit as the most obvious grouping within a home range area of Soay sheep on St Kilda island. Grubb (1974a) reported that: 'it was usual for the lamb of the year to keep close to its dam, and often a ewe yearling born to the dam in the previous year, accompanied them: female offspring more than two years old did not usually do so although they remained within the home range group'.

These family groupings suggest that the maintenance of these home ranges is probably a 'learnt' behaviour passed on from one generation to the next. Most reports seem to indicate that these groups are very stable over a number of years. In the Soay and mountain sheep ewe, home ranges are comprised of a number of groups in which females pass on the home range traditions to their offspring. This is further supported by Geist's (1971) observation that females rarely adopt home ranges other than those

Table 2.7. Home ranges and the number of ram mating contacts.

Home-range group	Ram			Number of sheep in home range
	Red	Blue	Yellow	
Bottom	16	33	15	38
Middle	20	44	21	52
Top	22	28	13	40
Fence	12	15	6	20
Total	70	120	55	150

Source: Adapted from Hunter (1964).

of their maternal band. It seems possible that lack of evidence for home-range behaviours in some sheep studies may be related to the age composition of the flock. Lack of previous experience of an area and/or the absence of older animals previously exposed remove the possibility of the self-perpetuating situations described above and consequently it is unlikely that home ranges are formed.

Males seem to establish less stable home-range areas compared with the ewes. The males split up into sub-groups and reform into new groups but extend their home range during the rut to encompass a number of ewe home ranges. In the study of Hunter (1964) three rams mated to ewes for a restricted period ranged across all ewe home-range groups (Table 2.7). Even within the ewe home range, young males have been observed to form

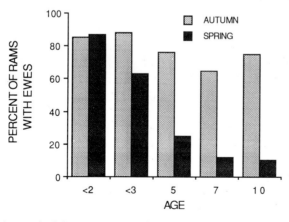

Fig. 2.17. The gradual disassociation with increasing age of Mountain Bighorn rams from the ewe flock (data from Geist, 1971).

sub-groups often moving with some autonomy from the main flock. Young males normally leave the maternal group as two-year-olds with the change being gradual (Fig. 2.17). Final establishment of a new home-range pattern in adult males may occur sometime after four and a half years of age.

The separation of males from the ewes and juvenile flock during a large proportion of the year may function to minimize competition for food during pregnancy and lactation. In both Rocky mountain sheep and in the Punjab Urials, rams and ewes remain apart for most of the non-breeding season; separation with the formation of male groups has also been reported in feral populations. Male groups were identified in all these studies with group size ranging from 1 to 20. In contrast Arnold *et al.* (1981) had difficulty in identifying separate male groups in 'feral' flocks of Dorset, Southdown and Merino sheep.

Breed differences

There is conflicting evidence concerning the existence of breed differences in expression of home ranges but it appears that home ranges are commonly observed in Longwool flocks in Britain. This is also reflected in the common usage of unfenced hill areas by different flocks which do not mix. Evidence of home ranges has also been reported for wild sheep and for feral domestic breeds.

In contrast, Lynch (1974) reported no evidence of home-range formation in Merino ewes grazing under extensive conditions in Australia. However, there was evidence of preferential grazing of certain areas related to distance from water and to a lesser extent to vegetation cover. The fact that the Merino ewe flock studied was relatively homogeneous in age structure may have contributed to the lack of an identifiable home range, with no tradition to pass from one generation to the next. Groups of Merino wethers which have been set-stocked over a number of years exhibit clear areas of preference in large paddocks.

Environmental effects on spatial organization

Social organization has to be flexible enough to adapt to changes in feed and social resources if a group of animals is to survive. Sub-group formation in sheep flocks is relatively common and would appear to be the result of both differences in breed cohesiveness and in environmental homogeneity. In situations where feed is sparse or where visual barriers are common we find sub-group formation is also likely. In fact sheep fit the generalized picture presented for ungulates by Leuthold (1977) '... changes in the distribution and abundance of food and water supplies largely determine the nature and extent of movement and with them the size of the home range of any individual, group or population of ungulates'.

Thus it seems that the overriding factor involved in spatial grouping of sheep is the flocking nature of the species. It also emphasizes the possibility that the experiences of the young when following the older sheep is the most likely method whereby 'traditions' such as home-range areas are established.

Social Organization and Behaviour in the Context of Handling

From the overview of social behaviour that has been presented, it seems that the process of domestication of sheep has not altered the basic behavioural patterns of the species. A number of characteristics of sheep have emerged, which have important implications for the handling and management of sheep. Kilgour (1976) suggests three basic behaviours which must be recognized for the successful handling of sheep. These are:

1. the strong flight reaction of sheep;
2. the dominant role of vision in social organization; and
3. the flocking/follower behaviour of sheep.

The way these basic behaviours influence the interaction of man with sheep flocks in the context of shepherding, yarding, pasture management, provision of limited resources of feed, water, shelter and confinement will now be examined.

Shepherding

Shepherding sheep has been practiced since the beginning of recorded history and is still widely used. However there is very little in the way of experimental data to quantify the methodology of the shepherd. Different shepherding methods are apparent in different countries, ranging from the training of animals to follow a person to the use of dogs to congregate and force the sheep to move. This fundamental difference in approach appears to be related to flock size, with small flocks commonly trained to follow the shepherd and large groups commonly controlled by a dog. As group size or dispersal of sheep increases so too does the signal intensity necessary to control movement. Consequently, as most leadership signals are either auditory or visual in nature, they are unlikely to be identified by the majority of animals in large groups. Movement is therefore controlled more by 'flock reaction' than by individual leaders.

Flocking

The flocking tendency of sheep is probably the most easily utilized behaviour to move or control sheep. Individual animals are almost impossible to

ALERT

FLIGHT

Fig. 2.18. The alternation of a flock between alert (facing the dog) and flight (moving away from the dog).

control as they panic when not in visual contact with other sheep. Most studies suggest a minimum of four or more sheep are necessary before predictable behavioural responses can be expected.

The alert response of sheep is normally activated by visual stimuli and this knowledge provides the shepherd with a number of means of controlling the animals. Sheep are extremely alert in the presence of man or dog and retain visual contact with them unless they turn to flee. This vigilance can be used by the shepherd to direct a flock in a certain direction or hold the flock steady in one place.

The flight distance of sheep varies considerably with a range from 5 m to over 1 km in wild Urial populations. Frequent handling of animals reduces this distance markedly. An awareness of this distance is fundamental to the manipulation of the flock from a point of being stationary (alert/attention) to slow movement (alarm) and finally to flight (Fig. 2.18). Flight distance also varies according to the area available for escape; wider races resulting in a greater flight distance (Fig. 2.19a). The general status of the flock in relation to flight can be monitored by observing the number of sheep facing the shepherd. The greater the proportion facing away, the closer the animals are to flight (Fig. 2.19b).

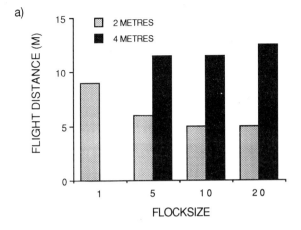

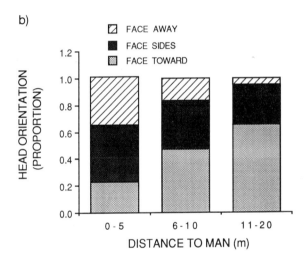

Fig. 2.19. The flight distance (m) in a laneway and the head orientation (proportion) of sheep as influenced by (a) flock size and (b) distance from man (data from Hutson, 1982).

The use of dogs for herding relies on the flight response of sheep to a 'predatory' species. The role of the domestic dog in this form of shepherding is vital and the behavioural characteristics of dogs which suit them for this task are summarized in Table 2.8 along with an indication of the possible origins of these behaviours in hunting ancestors. In a study of stimuli influencing heart rate in mountain sheep, it was shown that one of the most fearful of stimuli to mountain sheep was a man and a dog together (MacArthur *et al.*, 1979).

The man and dog combination is often used in the flocking management

Table 2.8. A comparison of the behavioural responses of the working dog with those of their predatory ancestors.

Working dog response	Predatory response
Alertness to individual strays	Strays selected as prey
Tendency to guard strays	Strays selected as prey
Able to immobilize sheep with 'eye'	Hunting position
Circle around flock to establish a position on the opposite side of the flock to the shepherd	Establish a position relative to the pack leader
Drive sheep to shepherd	Move prey toward pack leader

Source: Adapted from Vines (1981).

of sheep. In such situations, animals are often maintained in an alternating state between attention on the shepherd, as the sheep move away from the dog, and alarm. This state allows the shepherd to direct animals simply by the extension of a hand or crook, the sheep moving away from the extension. Sound can also be used, loud and abrupt sounds usually causing the animals to flee, soft and more prolonged sounds causing alert and relative immobility of the flock. It is difficult to determine whether there are breed differences in ease of handling as it is rare for a number of breeds to be reared under the same conditions and consequently have had similar opportunity to habituate to shepherding procedure. A study by Whateley *et al.* (1974) compared a number of breeds which had been reared together (Romney, Border Leicester × Romney, Perendale, Cheviot, Corriedale, Romney × Dorset, Romney × Merino sheep). These breeds were assessed for ease of mustering and reaction to dogs. Perendale, Cheviot and Merino cross animals were identified as the easiest to handle in such situations while Romney-Cross animals were the most difficult. These breed differences suggest that an awareness of particular breed characteristics is necessary in determining how easily animals can be managed and the method used in managing them.

The tendency for flocks of sheep to circle can also be used to the shepherd's advantage. This behaviour is the result of the greater speed of movement of animals closest to the source of 'fear'. The shepherd can manipulate the direction of flock movement by being positioned in such a way as to allow the animals to circle away (Fig. 2.20). This circling characteristic is evident in the more gregarious breeds and the impact of greater dispersal of some breeds on this pattern has not been reported.

An original application of the flocking tendency of sheep in response to

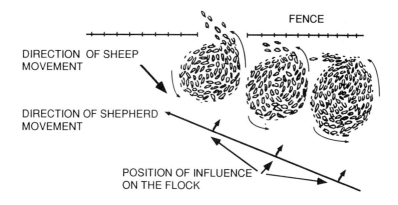

Fig. 2.20. The circling motion of sheep and the influence of a shepherd's position on the direction of movement.

predators has been made by Hulet *et al.* (1987) and Anderson *et al.* (1988) in rangeland conditions. In this environment, coyote predation is a problem but by rearing sheep and cattle together a new flocking structure is established. At the appearance of a coyote, 'bonded' sheep disperse themselves among the cattle rather than forming a separate flock, thus providing themselves with the protection offered by the cattle.

Following

The other characteristic of sheep which has been used by shepherds for thousands of years is that of following. The practice is clearly portrayed in biblical images such as that in the 23rd Psalm where the shepherd led his flock. The tendency of sheep to follow a leader is still used in many societies as a means of controlling the direction of movement of sheep. A trained animal, often a wether or goat with a bell around its neck, is used as a leader. These animals are trained to follow a shepherd from one pasture to another with dogs often being used at the rear of the flock to stop the animals dispersing. In some husbandry systems, dogs are used more in this way (guarding) than as working (driving) dogs, a use apparently only commenced in the mid-19th century in Britain.

Yarding

There are two major techniques which can be used to move sheep in yards. These are pushing by the use of fear and flight or the use of the follower-leadership pattern.

The basic behavioural patterns of sheep can be used to advantage in handling. The normally gregarious sheep are forced into a situation of reduced social space and quite likely, if they are pushed into a narrow race, into a state of semi-isolation. Such social disruption tends to result in an increased likelihood of flight particularly where the animal becomes confused by a number of other negative stimuli such as dogs, noise or an unfamiliar environment. In such circumstances sheep will normally seek to return to the flock either by turning or by charging forward.

Making use of flight behaviour appears to be an appropriate method for movement of sheep in yards as long as there is an obvious escape route. Without this route a number of major difficulties in the use of force will occur with animals constantly balking before reaching the desired point in a raceway.

The role of the senses

The importance of visual stimuli must be remembered when handling sheep in yards. Sheep need to see other sheep and, preferably, the latter should be moving. The design of yards should maintain races wide enough for a number of animals to be side by side in parallel races or, in the case of races which are only one sheep wide, should provide a curved race allowing animals to see the disappearing animal in front. Such a design allows animals to retain visual contact with other sheep and avoids isolation and consequential unpredictable fearful behaviour.

Sheep balk at changes in light or shadow particularly if there is bright light under floor grating (Hutson, 1981). The good depth perception of the sheep initiates an innate locking of front legs as a protection against falling thus causing a halt to movement. This effect is most notable when slats are

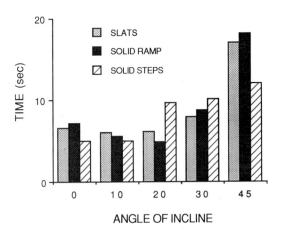

Fig. 2.21. The time (sec) for the first sheep of a flock to move up a 1.5 m wide ramp at different angles and with different surfaces (data from Hitchcock and Hutson, 1979).

Table 2.9. Means for latency and time taken by the first sheep to move from the start to the 3 m point in a race.

Attraction of decoy model sheep

Direction of model	Head position	Latency (sec)	Time taken (sec)
Away	Up	14.1	7.1
	Down	7.4	5.8
Side-on	Up	18.6	4.9
	Down	45.7	27.3
Toward	Up	14.1	12.2
	Down	30.9	26.0
No-Decoy		20.9	18.5

Attraction of calls associated with various situations

Call	Latency (sec)	Time (sec)
Isolated	13.7	3.3
Sight of food	10.7	2.9
Ewe separated from lamb	19.2	3.3
Lamb separated from ewe	25.4	4.1

Source: Adapted from Franklin and Hutson (1982b,c).

aligned in the direction of movement rather than across. Sheep move more easily into light rather than into dark areas, the delay in movement being caused mainly by the leading sheep. Rail width and positioning must ensure avoidance of visual distractions such as shadows or the sight of other sheep to the sides.

Merino sheep move faster on flat rather than on sloping surfaces and up rather than down gradients (Fig. 2.21). Therefore inclines should be gradual if they are to be used to move animals into sheds or to load sheep into trucks. Sheep tend to investigate, and therefore balk at changes in texture of surfaces underfoot.

In a series of experiments on Merino ewes, Franklin and Hutson (1982a,b,c) identified visual stimuli such as decoy sheep, mirrors at the end of raceways and photos as the most effective attractants to sheep. All stimuli attracted sheep into a laneway. There was some evidence that movement and posture may have been important as a model decoy was not as attractive to the sheep as live animals or photos, particularly if the decoy was facing or side-on to the flock and with head down (Table 2.9). Such a position may have been perceived as an aggressive stance by approaching animals. Mirrors also seemed to slow movement as animals are hesitant to move past the mirror and the 'animal' facing them. On the other hand a rump view has been identified as a 'follow me' signal in mammalian species, and this appears to be the case for sheep.

Olfactory stimuli had no effect on sheep movement and auditory stimuli only minor effects. Vocalizations associated with isolation or as a response to food had a greater attraction than those associated with ewe/lamb communication (Table 2.9).

The use of training to aid the flow of sheep through yards has been confirmed by Hutson (1980). Previous experience of yards and the same sequence of movement has a positive effect on flow rates of sheep. This suggests two things:

1. that animals should always follow the same pathway through yards; and
2. that the same yard design should be used on properties where there are several sets of yards.

Reinforcement of this pattern using food rewards is likely to have positive effects on ease of movement (Hutson, 1985) but sheep will recall aversive experiences in the same way. Consequently activities which are likely to be aversive such as dipping and shearing should be carried out in areas which are diverted from the routine 'animal movement' areas. Research by Hutson and his colleagues make it clear that the major delays in movement of sheep in yards is often associated with the first animal. Once this animal is moving the remainder of the group usually moves quickly unless there is a major design fault. Therefore, it seems that some

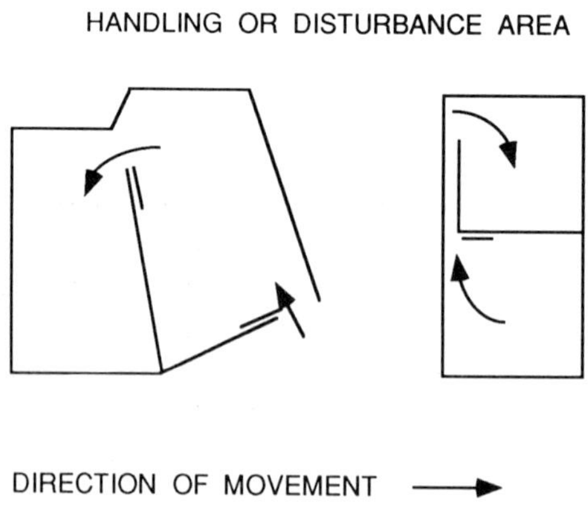

HANDLING OR DISTURBANCE AREA

DIRECTION OF MOVEMENT ⟶

Fig. 2.22. Designs of catching pens where animals move back toward other animals and away from handling/disturbance areas.

Table 2.10. Breed ranking (1 best–9 worst) for speed of movement through a series of yards and the mean time taken to catch sheep for shearing.

Breed	Flat yard	Steep yard	Catching time (sec)
Romney	5.5	8.0	10.7
Border Leicester × Romney	5.5	7.0	10.5
Dorset × Romney	3.0	6.0	10.2
Merino × Romney	4.0	3.0	9.8
Drysdale × Romney	9.0	9.0	9.7
Corriedale	8.0	4.0	11.5
Perendale	7.0	5.0	10.6
Cheviot	2.0	1.0	10.3
Merino	1.0	2.0	10.8

Source: Whateley *et al.* (1974).

form of decoy or trained leader is probably the best way to move sheep in yards. Descriptions of the successful use of 'Judas' animals to lead sheep in killing works has been reported (Bremner *et al.*, 1980).

In situations such as catching pens where animals are not required to continue movement, mirrors may be useful in attracting sheep to move forward and into the pens. Decoy pens of animals use excess space and are not the most desirable alternative in such situations. Alternatively, semi-circular pen or race designs may assist in such movement (Fig. 2.22) with animals always moving to open space or back toward other animals rather than towards a blank wall.

Breed differences

Breed differences in ease of handling in yards, filling catching pens, drafting and holding have been reported. The breeds easiest to handle in terms of mustering and yard movement were usually the most difficult to catch and physically restrain. Table 2.10 shows the ranking of various breeds for speed of movement of 80 ewes in full wool through a complex yard layout and the time taken to restrain these breeds (Whateley *et al.*, 1974). These data suggest that it may be necessary to alter yard handling procedures according to breed. They also imply that the more gregarious breeds are the easiest to move in yards, possibly due to their greater motivation to 'follow' but further studies are necessary to confirm this.

Innumerable yard designs have been published, and it is beyond the scope of this book to evaluate these. However, it has been noted that the circling nature of some sheep breeds can be used to advantage. Circular, or

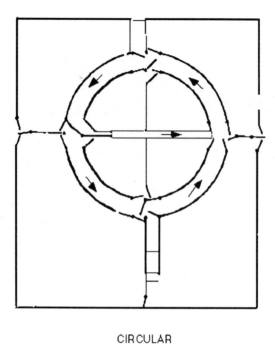

CIRCULAR

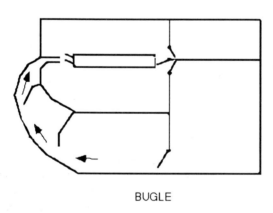

BUGLE

Fig. 2.23. Yard designs used for large flocks and utilizing the following and circling nature of sheep.

Fig. 2.24. The curved funnel leading to the race of a circular set of yards.

semi-circular yards with operators situated centrally allow for the natural tendency of circular flow of sheep and this has led to possibly the most popular yard design features in Australian yards (Fig. 2.23) with the use of a curved funnel leading into the race (Fig. 2.24). Circling movement is also an important consideration when positioning gates, and in determining the direction in which gates swing; always in the same direction as the normal circling motion. Yard design requires knowledge of the principles already discussed plus a measure of patience as minor modifications are usually required before sheep flow easily and quickly in yards.

Pasture and range management

The dominant feature of the social behaviour of sheep relevant to pasture management is flocking. This pattern is predominantly expressed when animals are resting or during times of disturbance. The flocking tendency results in the formation of resting sites of flocks in adjacent paddocks often being next to one another. It may also mean a very uneven distribution of nutrients particularly in paddocks where animals are set-stocked. The latter has implications for pasture utilization (see Chapter 1) and must be considered when fence positions for subdivisions are planned. The potential effect of home-range behaviour is important in the context of the design of paddocks, but this is not normally a problem in husbanded flocks where the restricted age range results in home ranges not being formed. However,

Fig. 2.25. Confinement of large numbers of ewes by an electric fence.

established home ranges may be used to reduce the need for subdivision with flocks remaining in appropriate home range areas without the need for fencing. This management alternative, where different flocks graze on communal areas without the flocks mixing, has been reported by a number of authors.

Breed differences in dispersal need to be considered when evaluating the likely utilization of a pasture and possibly even in determining optimal stocking rates. Arnold and Dudzinski (1978) have illustrated breed differences in dispersal in small 1 ha paddocks, with a threefold difference in area occupied by different breeds. Consequently, one would predict a more even or efficient utilization of heterogeneous pasture by those breeds with greater dispersal and it is possibly this characteristic in particular which has determined the division of breeds into mountain, hill and lowland types; the mountain and hill breeds tending to disperse evenly over large areas. The 'trade off' between ease of mustering and efficiency of pasture utilization has tended to favour mustering ease as the most desired characteristic in the expanses of Australian rangelands.

Habituation to electric fences has been used effectively as a means of intensive utilization of pastures in highly productive areas. Normal management requires high concentrations of sheep in small areas and the use of electric fencing is a cheap means of controlling sheep in this situation (Fig. 2.25). The electric fence relies on the fact that sheep will change their behaviour as a result of exposure to an electric charge. Retreat from the fence is the desired response and this usually occurs in sheep flocks after one exposure and subsequently infrequent reinforcement is necessary. It seems that there may be some social facilitation in the learning process as training of a few animals appears to affect the flock as a whole.

Provision of limited resources of feed, shade and water

There is no clear evidence that dominance status in sheep is related to priority access to feed, although limited access to feed increases levels of aggression (Arnold and Maller, 1974). Dominance may be of some importance in situations where supplementary feeding is practised. In a study by Sherwin and Johnson (1987) dominance in wethers, measured as a priority access to feed, was related to access to shade particularly during high ambient temperatures.

The consequence of increased crowding appears to be increased competition for resources. In the case of sheep, this is usually observed as an increase in the number of animals excluded from the resource; particularly if the resource is limited such as might be the case with supplementary feed. In trials using Border Leicester × Merino ewes, the progressive reduction of trough space from 24 to 4 cm per animal led to an increase in the number of ewes failing to eat, from 0 to 31%.

There is evidence of a direct relationship between dominance rank and mating performance in multi-sire matings. In field studies, dominance is also related to mating success with a 28% reduction in flock fertility being observed in one study where the dominant ram was vasectomized (Fowler and Jenkins, 1976). The role of dominance in mating behaviour under different conditions will be discussed further in Chapter 3.

Confinement

Density is a spatial concept while crowding implies a large number of animals as well as restricted space and can be defined in terms of invasion of personal space resulting in increased levels of aggressive encounters. Crowding is difficult to evaluate as it depends on either excess density of animals or limited resources for the animals present. Both situations can cause competition, particularly the latter.

Crowding of sheep occurs at feed or water troughs, during intensive 'strip grazing' situations where upwards of 3000 animals may be in areas of less than 1 ha and in housed groups. Breeds differ in their responses to this management situation. The Merino tends to remain flocked and utilize only a limited quantity of feed compared with the lowland breeds who disperse themselves evenly over the area and utilize all the feed available.

There is little experimental evidence to show how stocking rate influences the social behaviour of sheep apart from suggestions that breeds differ in their responses to herding, handling and confinement. 'Open flocking' breeds such as the Scottish Blackface are individualistic and easily frightened when crowded together. More close-flocking breeds such as the Romney appear to suffer less 'stress' in confined or housed situations.

Hutson (1984) has examined spacing of animals in group pens and has noted that sheep tend to be relatively evenly spaced in the pen when standing but in parallel orientation to one another, avoiding head to head confrontation. The animals were arranged around the periphery while lying. These observations suggest that spatial requirements do exist in confinement and that some care must be taken to provide adequate space for animals to orientate themselves appropriately. There appear to be no definitive studies on this aspect of sheep behaviour.

Sheep in confinement may exhibit abnormal behaviours. In particular individually housed animals can exhibit stereotypies, repeated actions without purpose, such as wool or pen chewing. These will be discussed further in Chapter 6.

A concept which has been used in intensive management of some domestic species is that husbandry practices tend to lead to a reduction in behavioural stimuli and in complexity. Kiley-Worthington (1977) reports that reductions in social complexity can be compensated for in rodent species by increasing the complexity of the physical environment. This suggestion has also been confirmed in a study of grazing Merinos (Stolba *et al.*, 1990). The implications of the maintenance of a relative complex social and/or physical environment may be the avoidance of abnormal behaviours including stereotypes in housed animals and also in the maintenance and expression of a complete behavioural repertoire.

Conclusions

The handling of sheep should be based on an understanding of the basic behavioural responses of these animals. In particular the flocking and follower characteristics can be used to advantage as can an understanding of the dominance of visual stimuli in initiating sheep responses.

In terms of the more general organizational characteristic of the sheep, Kilgour's (1972) suggestion that the greatest influence man has on adult animals is the way he allows social relationships to develop may well be true. At present, social relationships are modified via sex or age segregation of sheep and the impact of this on animal distribution and grazing patterns needs further evaluation. We may find that the modification of social relationships is not always beneficial to productivity.

Further Reading

Arnold, G.W. (1985) Association and social behaviour. In: Fraser, A.F. (ed.), *Ethology of Farm Animals*. Elsevier, Amsterdam, pp. 233–46.

Craig, J.V. (1986) Measuring social behaviour: social dominance. *Journal of Animal Science* 62, 1120–9.

Hunter, R.F. (1964) Home range behaviour in hill sheep. In: Crisp, D.J. (ed.), *Grazing in Terrestrial and Marine Environments*. Blackwell, London, pp. 155–71.

Geist, V. (1971) *Mountain Sheep – A Study in Behaviour and Evolution*. University of Chicago Press, Chicago.

Grubb, P. (1974) Social organization of Soay sheep and the behaviour of ewes and lambs. In: Jewell, P.A., Milner, C. and Morton-Boyd, J. (eds), *Island Survivors: The Ecology of the Soay Sheep of St Kilda*. Athlone Press, London, pp. 131–59.

Schaller, G.B. (1977) *The Mountain Monarchs: Wild Sheep and Goats in the Himalayas*. University of Chicago Press, Chicago.

Syme, G.J. and Syme, L.A. (1979) *Social Structure in Farm Animals*. Elsevier, Amsterdam.

3

The Reproductive Behaviour
of Sheep

Introduction

There is a surprising commonality in patterns of mating behaviour across the wide variety of ovine species and breeds. The *Ovis* species is recognized as promiscuous with no evidence of the development of pair bonds between male and female. It is this promiscuous mating pattern, along with flocking behaviour, short flight distances and precocial young which may have led to the domestication of this species. These characteristics have been retained in the wide diversity of breeds now in existence and the mating pattern remains essentially unchanged. The initiation of mating behaviour is largely controlled by a 'seasonal switch' which allows the expression of oestrus in the ewe which in turn triggers the expression of libido by the ram. These sexual behavioural drives are under endocrine control with changes in hypothalamic, pituitary and steroid hormones determining the cyclicity of oestrus in the ewe and the presence or absence of libido in the ram.

The seasonality of breeding in feral and wild populations of sheep is associated with the segregation of the sexes for much of the year. Segregation has also been enforced by most of our modern husbandry practices by exclusion of rams from the ewe flock or by castration. However, the early exclusion of rams from the ewe flock in most intensively husbanded flocks is not a characteristic of wild populations where young rams remain with the ewes for a prolonged period. This change in social environment has a number of implications for the mating behaviour of rams used in intensive management systems.

This chapter will examine the various aspects of the mating behaviour of the sheep and will highlight behavioural characteristics of particular

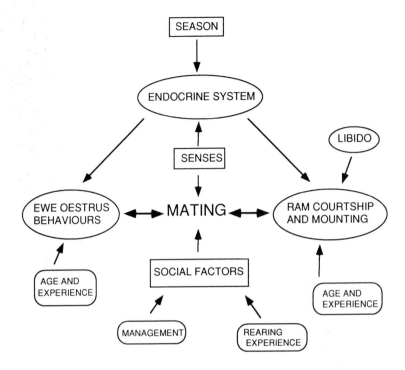

Fig. 3.1. Factors influencing the occurrence of reproductive behaviours in the sheep.

importance to the management of sheep flocks. No attempt will be made to separate the behavioural descriptions of male and female since, to a large extent, these behaviours are interdependent. The model presented in Fig. 3.1 illustrates the links between the two sets of behaviours and summarizes the contents of this chapter.

Sexual Drives

Mating behaviour can be seen as the behavioural sequences established to allow the expression of male and female sex drives and allow mating to take place.

Oestrus is the term used to describe the period of sexual receptivity in the female. It refers to a period of high activity and excitability in many species but this is not normally the case for sheep. The ewe usually goes through a series of oestrous periods during a breeding season and may have at least one silent ovulation (ovulation without behavioural oestrus) at the commencement of the breeding season.

The ewe is undemonstrative when in oestrus and it is difficult to identify oestrus by behavioural characteristics except those associated with ram responses. Ewes usually remain in oestrus for about 40 hours but this varies with breed and age, and ranges from three hours to three days have been reported. There is evidence that repeated contact with rams will shorten the duration of oestrus and a difference of around ten hours between ewes intermittently exposed to a ram and those continuously exposed to rams is common.

Sexual drive of rams is now commonly described as libido. This term was first used by Freud in the context of sexual instinct represented in the mind but has now been accepted as a term to describe the general level of sexual drive in domestic animals. Wodzicka-Tomaszewska *et al.* (1981) have reviewed the various definitions of the term and its use in the context of assessment of male sexual behaviour.

The libido of rams is expressed during the period of ewe oestrus by a series of courtship behaviours leading to mounting and intromission. The libido of rams is usually assessed in relation to the number of ewes with which a ram can mate in a set time. This definition relates particularly to intensive ram use. Measures of courtship and dominance should also be considered within the context of seasonality and the segregation of male and female groups. Before we examine these effects, the role of the endocrine system in control of mating behaviour will be examined.

Endocrine Mechanisms Controlling Mating Behaviour

The ewe

The reproductive behaviour of ewes is strongly influenced by hormonal effects (Fig. 3.2). In the 1950s it was shown that the presence of the ovary was essential for the exhibition of oestrous behaviour which was cyclic in nature and occurred at approximately 17-day intervals during the breeding season (Robinson, 1955).

More recently, follicle stimulating hormone (FSH) and luteinizing hormone (LH) have been identified as the two main gonadotrophic hormones secreted by the pituitary gland which control the oestrous cycle. FSH stimulates the growth of ovarian follicles which secrete oestrogens associated with the onset of oestrus. Under the influence of pulsatile LH secretions, ovulation will occur after the regression of the corpus luteum. Within 24 hours behavioural oestrus is observed in response to a large release of oestrogen. Subsequently, a new corpus luteum forms and the ewe remains sexually unreceptive during the period when progesterone secreted from the corpus luteum is the dominant steroid.

Oestrogen is the predominant hormone involved in the expression of

THE EWE   THE RAM

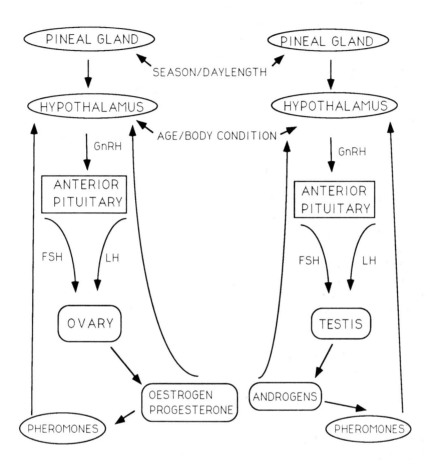

Fig. 3.2. The endocrine mechanisms influencing reproductive behaviour of ewes and rams.

oestrus but animals are more sensitive to its action after exposure to progesterone. Progesterone appears to prime the central nervous system for oestrous behaviour and this 'priming' increases the intensity of oestrus, as measured by the intensity of courtship behaviours (Fig. 3.3) exhibited by rams during the oestrous period. However, progesterone must be at a low level at the time of ovulation. The ewe is also more sensitive to oestrogens during the breeding season.

The length of behavioural oestrus is dose-responsive to oestrogens in ovariectomized ewes but the length of oestrus is shorter than in natural cycles (12 vs. 30 hours in mature ewes). Time to onset of oestrus after

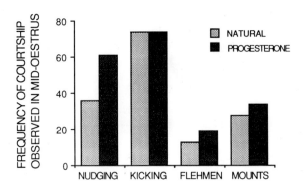

Fig. 3.3 The frequency of nudging, kicking, flehmen and mounts exhibited by rams toward ewes in natural oestrus or in a progesterone 'primed' oestrus (data from Tomkins and Bryant, 1974).

treatment is also dose-dependent with higher doses inducing earlier onset (30 vs. 25 hours post-injection).

Other steroids such as testosterone and androstenedione can also induce oestrous activity at high doses, while high doses of oestrogen may also increase aggression in the ewe. A more detailed review of the endocrine basis of female reproduction has been represented by Signoret (1980).

Periods of transition: season and puberty

The seasonality of the breeding season in sheep is due to the declining day length which increases the secretion of melatonin from the pineal gland. In turn the level of melatonin influences the production of gonadotrophin releasing hormone (GnRH) in the hypothalmus which controls the release of the gonadotrophins LH and FSH from the pituitary (Fig. 3.2). These latter hormones are necessary for the normal functioning of the ovary both in producing the steroids necessary to initiate oestrous behaviour and to allow ovulation. The breed variation in the onset of the breeding season appears to be determined by the depth of anoestrus, some temperate breeds requiring a longer period of melatonin stimulation to initiate GnRH secretions.

For those breeds, including the Merino and most breeds of tropical origin, which do not have a great depth of anoestrus another phenomenon can also initiate oestrous activity. The 'ram effect' has been recognized for many years and was first documented in Merinos in the 1950s (Schinkel, 1954). Ewes are stimulated by the unaccustomed presence of rams, or testosterone-treated wethers, to initiate GnRH secretion and thus increase tonic secretions of LH. This in turn usually results in ovulation two or three days after ram introduction.

The length of the oestrus cycles initiated in this way may be the normal

17 days but in some cases short six-day cycles can occur initially. The latter is associated with lack of progesterone priming but an exogenous source of progesterone may be used to ensure that normal cycling patterns are established.

The ram

Sexual activity in rams is under the control of the androgenic hormones produced by the testis (Fig. 3.2). If rams are castrated after puberty they lose sexual activity slowly. Adult castrates retain some sexual activity for up to 12 months. It is apparent that libido is not related to the level of male hormones and a number of studies have noted that circulating levels of testosterone in the male ruminant are normally well above the threshold needed to elicit sexual behaviour in the ram. It seems likely that other factors are largely the cause of variation in libido.

Therapeutic doses of testosterone or LH have not been shown to improve libido in rams although steroids induce the onset of male libido in wethers. Treated wethers have been reported to exhibit male sexual behaviours within ten days of the commencement of testosterone therapy and animals continue to exhibit male behaviours for four to seven weeks after cessation of testosterone therapy. This may vary depending on the androgen esters used in treatment.

Recent data suggests there may be differences in the pattern of LH release or response to luteinizing hormone releasing hormone (LHRH) in high and low libido rams and this link between gonadotrophins and behaviour is further supported by seasonal patterns. Seasonal variation in sexual activity of rams has been reported with activity being lower during the

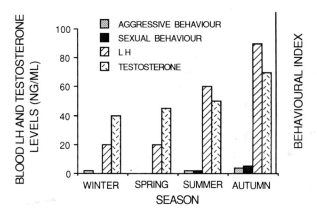

Fig. 3.4. Seasonal changes in testosterone, luteinizing hormone and aggressive and sexual behaviours of Soay rams (data from Lincoln and Davidson, 1977).

spring and early summer when daylight is increasing and FSH and LH levels are lower. Consequently testosterone production also tends to be lower at this time.

This pattern is most noticeable in the more seasonal breeds. For example, Fig. 3.4 illustrates the parallels between endocrine changes and sexual behaviour. Even in less seasonal breeds such as the Merino, libido also seems to wane during the spring both in rams and testosterone-treated wethers. The latter effect may be associated with a changing sensitivity to testosterone and parallels the sensitivity to oestrogen already described for the ewe.

Endocrine effects on sexual differentiation

Sexual differentiation of behavioural patterns in domestic livestock has been reviewed by Ford and D'Occhio (1989). They reported that these patterns are established as a result of the actions of steroid hormones on the brain during discrete sensitive periods of fetal development. The mammalian brain is inherently female and, in the male, changes are related to the defeminization and masculinization of sexual behaviours which are the culmination of a series of exposures to testicular steroids. In the ram, defeminization of the brain occurs between days 50 to 80 of gestation. This is well illustrated by the fact that fetal ewes exposed to testosterone between days 50 to 80 fail to show sexual receptivity and exhibit aggressive and mounting behaviours as adults.

A post-natal phase of sexual differentiation has also been proposed for the ram. Testosterone supplementation between four and eight weeks of age appears to increase the libido of rams as adults. The reverse is also true. Rams receiving antisera to testosterone at this time have a reduced libido as adults. However, the pre-natal phase is clearly of greater importance as rams castrated neonatally will exhibit normal sexual behaviours when given androgens as adults.

Seasonality and Segregation of Sexes

Evidence from a range of sheep species suggests that there is a segregation of sexes for much of the year. Schaller and Mirza (1974) report 'the rams showed little interest in ewes during early October and the only evidence of the approaching rut was the occasionally roaming ram'. This reflects a seasonality of breeding season the onset of which is controlled by declining day length. In temperate zones, sheep are seasonally polyoestrous with mating in autumn and a summer anoestrus. However, there are exceptions to this pattern, particularly for breeds of tropical origins which can breed for a large proportion if not the whole year.

Feral and wild populations

In Soay sheep (Grubb, 1974b), the rut commences in early autumn (October) with the breakdown and dispersal of male groups and the establishment of wandering males 'scanning' for oestrous females. A similar timing is also reported for the wild Mouflon. Males are often seen in conflict during this time but the incidence of this declines toward the end of October when most ewes are in oestrus.

In mountain sheep, the rut commences in late autumn but is preceded by a period when the rams gather together before dispersing to set rutting areas. During the period of pre-rut the rams often display and form huddles which Geist (1971) suggests help to stabilize dominance rank among rams. It is during this time that most of the severe fights are observed, particularly among large-horned older rams.

Once rams have dispersed to rutting areas, individual rams can be seen moving through the flock seeking oestrous ewes. Rams will approach ewes in a low-stretch 'threat' posture, an initial behaviour of courtship. Anoeostrous ewes tend to flee while ewes showing early signs of oestrus will be 'guarded' by the ram until they become receptive. The behaviours which follow seem to be similar for all groups of sheep.

The onset of oestrus in the ewe varies by a few weeks from year to year and this may be associated with nutritional status. The rut is usually spread over two or three weeks in wild populations of mountain sheep, Soay and Mouflon, but the rams may stay with the ewes for up to eight weeks.

Once the rut or breeding season has finished, rams re-establish a ram group and move away from the ewe home-range areas. However, rams of some breeds return to or remain with the ewe group. For example feral populations of 'Merino type' living on Campbell Island of New Zealand have lambs born over a six-month period (Wilson and Orwin, 1964).

Breed variation in the length of the breeding season is illustrated by the markedly different patterns of lambing of four breeds run under identical conditions of minimal husbandry in south-western Australia (Fig. 3.5). All breeds had some seasonal peak but the Merino and Dorst Horn rams remain with the ewes throughout the year and the distinct breeding season pattern of the wild species does not occur.

The possibility that the segregation of males and females could result in a 'ram effect' with synchronization of oestrus in feral and wild sheep has been suggested (Lindsay, 1988). This hypothesis is supported by the deliberate separation of male and females for a two- or three-month period pre-rut in the wild populations mentioned earlier. However, young males up to two years old are present with the ewe flock in most wild sheep populations and consequently it can be argued that it is not only the ram's presence and the effect of pheromones that initiates the onset of oestrus but also the intense activity of initial ram–ewe contact which is stimulatory. Before

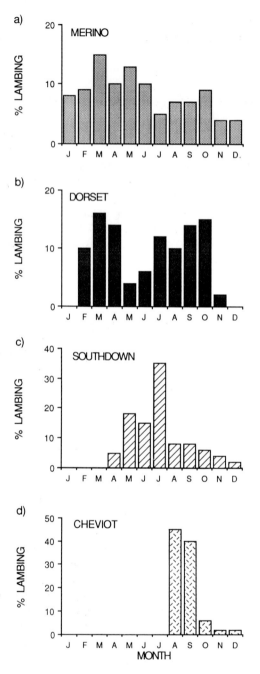

Fig. 3.5. Seasonal patterns of lambing for populations of: (a) merino; (b) Dorset Horn; (c) Southdown; and (d) Cheviot ewes in south-western Australia (reproduced with permission from CSIRO, 1982).

examining the behavioural sequence of mating, the role of the various senses in broadcasting sexual state and in stimulating sexual responses will be examined.

The Role of the Senses in Mating Behaviour

Reproductive signals have been previously categorized into those for attracting the opposite sex (broadcasting), those which demonstrate dominance, those which stimulate oestrus in the female and those which identify the stage of oestrus or stimulate the male. The major way in which these signals are communicated in the sheep appear to vary. Broadcasting and dominance are associated with visual cues while sexual state and stimulatory signals seem to be predominantly olfactory. Auditory stimuli, although present on occasion, do not appear to play a major part in the repertoire of reproductive behaviour in the sheep.

Vision

Visually impaired rams have difficulty locating ewes even when ewes are able to move to the rams. When the senses of sight smell and hearing of ewes are impaired mating success is reduced (70 vs. 55%), particularly if sight is impaired and rams are tethered. This implies that the oestrous ewe has a major role in identifying and moving to the ram.

Olfaction

The use of scents is important in the reproductive processes of ungulate species. Scent marking is usually linked with agonistic rather than reproductive behaviour. Courtship behaviours often involve aggression and there is no clear-cut distinction between the two. As mentioned earlier (Chapter 2), it is difficult to differentiate between the olfactory stimuli used for reproductive purposes and those which function in maintaining social organization. It is not unusual to find that postures used in threat or dominance also appear in the sequences of mating behaviour in ungulates.

Most ungulate males determine oestrus by investigation of the anogenital region or the urine of females and, in sheep, this behaviour includes sniffing, licking, nuzzling, face rubbing, and possibly nibbling movements with the lips. Flehmen (curling of the upper lip) is a behaviour thought to be involved in the function of the vomeronasal organ which has receptors particularly sensitive to low-volatile compounds such as oestrogens. Flehmen is exhibited in response to several odours including those from urine of the oestrous female and amniotic fluid. It is most frequently exhibited by the ram in investigating ewes during 'courtship' and probably reflects the identification of odours from an oestrous ewe.

The ability of rams to discriminate between urine from oestrous and non-oestrous ewes was tested in an operant conditioning study where it was shown that rams could repeatedly distinguish the urine of oestrous ewes (Blisset *et al.*, 1990). Rams in this study did not exhibit flehmen which implies that flehmen is not necessary in scent identification and may play some other role in courtship.

The role of olfaction in the ability of rams to discriminate between oestrous and non-oestrous ewes was demonstrated when anosmic rams were shown to approach both oestrous and non-oestrous ewes at random; they only identified ewes in oestrous by attempting to mount (Table 3.1).

The ewe can be influenced by the pheromones produced by the ram. There is now evidence that the smell of rams or their wool (Knight, 1983) can induce oestrous in anovular ewes. This effect is also apparent if oestrogen- or testosterone-treated wethers are used instead of rams. There are also differences between ram breeds in their effectiveness as 'teasers'. In one study, Dorset Horn rams induced a three-week earlier onset to the breeding season compared with Romney or Finn × Romney rams (Fig. 3.6).

The role of experience may be important in determining the effectiveness of the senses as there is ample evidence that inexperienced rams running with adult ewes may take weeks before they exhibit normal sexual behaviour.

Rams exhibit a preference for their own breed of ewe when offered a choice. From studies of cross-fostered rams it appears that this preference is related to the breed of dam, that is rams prefer to mate with the same breed as their mother or foster mother. However, there is no conclusive data available to indicate which of the senses may be involved in recognizing the breed of the ewe.

Table 3.1. Oestrous ewes served by rams when rams or ewes were sensory impaired (%).

	Rams impaired		Ewes impaired	
Impairment	Tethered ewes	Free ewes	Tethered rams	Free rams
Control	96	93	68	85
Deaf	86	87	–	–
Anosmic	55	90	–	–
Blindfolded	48	50	28	80

Source: Adapted from Fletcher and Lindsay (1968).

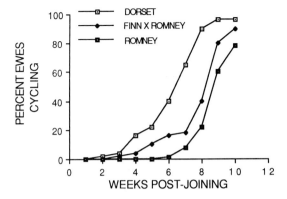

Fig. 3.6. The cumulative proportions of Romney ewes in oestrus at the commencement of the breeding season grouped according to the ram breed to which they were mated (data from Meyer, 1979).

The factors determining sexual 'attractiveness' of ewes has been reported by Tilbrook (1987a,b). Attractiveness was measured using a test consisting of recording time spent in courting and mating by each of six rams when given a choice of six oestrous ewes. It was shown that:

1. attractiveness was not related to level of oestrogen used to induce oestrous;
2. attractiveness was closely linked with oestrus and was relatively stable for an individual animal between and within oestrus periods;
3. attractiveness was a characteristic of an individual ewe and was not associated with soliciting behaviours;
4. attractiveness was independent of whether a ewe had been previously exposed to another ram;
5. attractiveness may be associated with wool characteristics as unshorn ewes were consistently more attractive to rams than were shorn animals.

The latter observation suggests the possibility of the involvement of scents on ewes' wool as already identified in the wool of rams. A later study confirmed that unshorn ewes were more attractive to rams than were shorn animals but only in the field (94.4% vs. 44.4% mated) and not in pens (Tilbrook and Cameron, 1989).

Vision appears to be the dominant sense used by both ram and ewe in making first contact, but olfactory mechanisms play a major role in identification of oestrus once initial contact has been made. Visual, olfactory and possibly auditory cues are all involved in the final behavioural sequence leading to intromission although the immobility of the oestrous ewe is of critical importance as a trigger for mounting.

Ewe–Ram Interactions and Mating Behaviour

Rams engage in a systematic search of oestrous ewes. The sequence begins with the ram approaching the ewe and if she does not move away the ram will nose her tail and vulva. The ewe may then crouch (squat) and urinate which is often followed by the ram sniffing the urine and exhibiting the flehmen response. If the ewe is in oestrous the courtship pattern is established with the ram following the ewe, in 'low-stretch position' and sometimes nuzzling the ewe's flank. Persistent tending of oestrous ewes has not been observed for Urial sheep but is clearly seen in Soay and Mountain sheep. In the latter, there are reports of a number of rams following a pre-oestrus ewe over long distances until she enters behavioural oestrus and stands for the dominant ram. This behaviour appears to be most common in habitats covered with dense shrubs where the ram must follow to retain visual links with the ewe.

In some cases, following activity may attract other rams and the continuance of courtship then depends on the dominance position of the courting ram. If the ewe is receptive then the ram will perform a series of courtship behaviours, the frequency and duration of which seems to vary widely between rams. Breed differences are also apparent with the frequency of nudging being far higher in Merino than in Suffolk rams while the frequency of kicks is higher in the Suffolk. Geist (1971) reports that Bighorn sheep display relatively less than thin-horned rams and are therefore more 'rough and unpolished' when courting ewes. However, the courtship sequence includes nudging the ewe before mounting. Some observers have noted that Bighorn ewes will only allow males, regardless of age, who perform complex courtship behaviours to mate with them.

The principal reproductive behaviours exhibited by the ram and ewe are summarized in Table 3.2. Ewes and rams do not exhibit all of the listed behaviours on every occasion. The data in Table 3.2 give some indication of the frequency of these behaviours during the final behavioural sequence leading to mating of mature Border Leicester × Merino ewes by Poll Dorset rams. Bighorn rams also exhibit the courtship behaviours described above and the frequency of these behaviours over the total courtship period is shown in Fig. 3.11. Nosing and twisting are the two most frequent courtship patterns. Kick (nudge) and twist are also the most frequent displays observed during inter-male interactions.

In a study of the Mouflon (McClelland, 1991) an evaluation of the behavioural combinations and components of behavioural sequences suggests that the behaviour of greatest frequency are different in male–male and male–female interactions. The kick (64%) and twist (28%) dominate male–male threat, while twist (46%) and flehmen (14%) are the predominant behaviours exhibited during male–female interactions.

Once the ewe stands for the ram she will usually tail-fan and turn her

Table 3.2. Description of reproductive behaviours of the ewe and ram.

	Frequency of occurrence of behaviour patterns of oestrus ewes and mature rams (%)
The ewe	
Squat/crouch: ewe in a crouching posture, usually includes urination and occurs after nudging by the ram	40
Circling: the ewe turns back toward the ram, often nuzzling his flank, ram usually follows to retain his position behind the ewe	20
Tail fanning: the ewe's tail is elevated and fanning in the presence of the ram (Fig. 3.7)	91
Head turning: ewe turns her head back toward the ram as he approaches to mount (Fig. 3.8)	81
Stand: ewe stands firmly when ram attempts to mount	97
Following/migration: the ewe follows a ram after initial contact by the ram, often in association with another ewe	55
The ram	
Sniff/nose: smell urine or perineal region of ewe (Fig. 3.9)	98
Flehmen: after sniffing the ram arches his head up and curls the upper lip showing his teeth	85
Low stretch: The low stretch involves the neck being held horizontal to the ground with the muzzle forward and raised. The head is often turned through 90° (twist) as well (Fig. 3.10)	85
Nudging: consists of one alone or a combination of the kick (forefeet used in a pawing motion while standing parallel to the ewe), rubbing (the ram rubs head and shoulders along or under the ewe's flank) and low stretch	96
Lick: licking ewe's flank, running tongue in and out	24
Mount: the ram's brisket in firm contact with ewe's rump	100
Ejaculation (service): pelvic thrust during mounting accompanied by a rapid backward movement of the head	100

Fig. 3.7. Positioning of a ram during tail fanning of the ewe.

Fig. 3.8. Head turning of an oestrous ewe toward a courting ram.

Fig. 3.9. Sniffing behaviour of a ram toward an oestrous ewe.

Fig. 3.10. The low-stretch posture of a Merino ram with head twist.

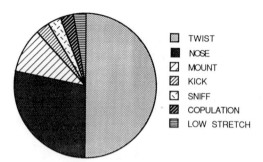

Fig. 3.11. The relative frequency of occurrence of sexual behaviours in mature mountain rams (data from Shackleton, 1991).

head back toward the ram while standing still. This behaviour is quickly followed by the ram mounting. Between one and four mounts usually occur before ejaculation. After ejaculation there is normally a refractory or latency period before the ram will again show interest in the same or another ewe. The period of latency can vary in length from a few minutes to over an hour. Ejaculation is usually identifiable by a deep pelvic thrust followed by throwing back of the head. It seems that there are breed differences as this behaviour is more clearly exhibited in Merinos than in Suffolk or Border Leicester rams and has not been observed in Mountain sheep or in the Mouflon.

After mating the ewe stays nearby to be served again, or she moves away to be mated by another ram, particularly if area and flock size allow the establishment of a reasonable spatial distance between rams.

An interesting variation on normal courtship behaviour has been reported by Kilgour and Winfield (1974). They observed that rams mating in steep hill country found it impossible to mate with ewes facing uphill. Butting was added to the nudging sequence prior to mounting so that the ewes were facing down hill!

Harem formation

There is general agreement from researchers that it is the ewe which holds the initiative for mating. The final determinant of mating is the ewe's response to ram courtship behaviours, with immobility, head turning and tail fanning indicating to the ram that she is receptive. Geist (1971) states that harems are not formed in mountain sheep and that the normal pattern is for rams to form a consort with individual oestrous ewes. This may be due to the small numbers of ewes in oestrus on any one day. Harem formation, in the sense of an aggregation of ewes around one ram, has been widely reported in large flocks and/or in oestrus-synchronized flocks of

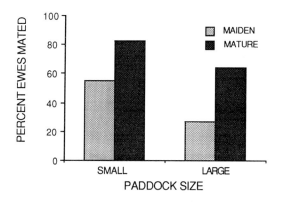

Fig. 3.12. The percentage of ewes mated in small and large fields when a tethered ram was used (data from Allison and Davis, 1976a).

domesticated sheep. Mattner *et al.* (1967) has described the formation of a harem of oestrous ewes around the dominant ram and noted that this did not exclude subordinate rams since ewes were observed to move in and out of the 'audience area' of the dominant ram.

Harem formation may be a consequence of the ewe courtship patterns rather than the gathering or herding of ewes by a ram. Ewes from domesticated breeds are known to seek out rams that are tethered (Lindsay, 1966). This is illustrated in Fig. 3.12 where the proportion of ewes mated by a tethered ram (Fig. 3.13) was lower for young than for mature ewes. Jewell and Grubb (1966) have also reported that mature Soay ewes will sometimes solicit rams.

Sexual Behaviours in Young Animals

Male lambs

Thwaites (1982) has detailed the ontogeny of mating behaviour in cross-bred ram lambs housed indoors. He reported an initial peak of mounting and courtship (teasing) behaviours in young rams at four to six weeks of age followed by a steady increase (Fig. 3.14) in the incidence of sexual behaviours until almost all rams exhibited courtship and mounting behaviours by 26 weeks of age. A reduction in the incidence of some courtship behaviours (lick and kick) was thought to be a result of the rams adapting their behaviour in response to the restraint of the teaser ewes which reduced the necessity for courtship prior to mounting.

The high incidence of mounting activity of young ram lambs has been noted in the field and appears to be associated with play activity. Male-like

Fig. 3.13. A tethered ram used in a series of field-mating behaviour studies.

play behaviours have been recorded in both ewe and ram lambs but only in the first month of life and not apparently associated with hormonal changes. The frequency of such play activity was higher in the absence of adult females.

Sexual behaviour of rams appears to develop in two phases: (i) the pre-pubertal period of intense sexual-type activity (play), and (ii) the period of adolescence when frequency of sexual activity is low but more adult-like. The functional significance of the first phase remains unknown as the frequency of infant sexual play is not related to the intensity or frequency

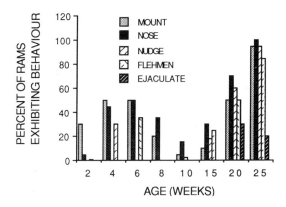

Fig. 3.14. Age changes in the incidence of male sexual behaviours in ram lambs (data from Thwaites, 1982).

of ewe behaviours. The second phase is clearly of importance as heterosexual contact at this time influences the sexual behaviour patterns of the ram.

Ewe lambs

Ewe lambs exhibit some male-like play behaviours in the first month of life but female sexual behaviours are not seen until the animal commences to cycle and is influenced by oestrogens. Inexperienced ewe lambs undergoing their first cycle exhibit an incomplete repertoire of sexual behaviours in response to experienced rams and a shortened behavioural oestrus. In one study (Kilgour and Edey, 1977) a third of the ewe lambs showed a full adult standing response, half showed tail fanning in response to nudging by the ram, a third showed head turning and 10% showed scrotal nuzzling (Table 3.3). Consequently, 12% of ewes were not inseminated by the rams even when mating occurred in pens.

Once the ewe has commenced to cycle regularly, the repertoire of sexual

Table 3.3. Ewe lamb courtship response patterns (% of total possible responses).

Tail fanning	%	Head turning	%	Scrotal nuzzling	%
Frequent	37	Full–frequent	28	With soliciting	10
Average	20	Full–rare	17	Nuzzling alone	15
Rare	25	Half-turn	17	Nil	75
Nil	18	Nil	38		

Source: Adapted from Kilgour and Edey (1977).

behaviours develops quickly with a complete behavioural repertoire exhibited by most ewes at their second or third oestrous cycle.

Rearing Experience and Ram Behaviour

Rams are usually run in all-male groups between weaning and their first mating experience at 18 to 24 months. It has been suggested that this practice can lead to high levels of homosexual or monosexual behaviour and a resultant lack of response to oestrous ewes. In some studies, 'non-working' rams were unresponsive to ewes in oestrous but were sexually responsive to other rams (Zenchak and Anderson, 1980). In one group of rams reared as a single-sex group only 50% responded sexually to oestrous ewes compared with 100% of rams reared in a similar manner but having opportunity to see, hear and smell ewes while remaining physically separated from them. Fowler (1984) reported that an average 27% of rams tested were found to be inactive on their first exposure to oestrous ewes. Many of these rams became active after repeated exposure to ewes but they tended to be 'below average' in their later mating performance. Rearing in isolation from ewes not only resulted in lower mating success (11 vs. 54%) but also delayed the onset of semen production in pubertal rams (18 vs. 50%) (Casteilla *et al.*, 1987).

It appears that a high proportion of rams reared in isolation from ewes do not recognize ewes as positive stimuli for sexual arousal and can be classified as 'non-workers'. This isolation does not occur in wild or feral populations where rams remain within the female group until they are more than 12 months old. The formation of all-male groups in the subsequent post-pubertal period is unlikely to contribute to an incidence of similar 'non-workers'. Mounting and other sexual behaviours between rams have been reported in wild populations but within the context of dominance encounters. The problem of 'non-working' rams which show inadequate sexual behaviours appears to accompany management practices imposed on domesticated sheep.

Mating Behaviour in The Field

A number of reports show a distinctly diurnal pattern of mating behaviour with peaks in the early morning and late afternoon. This may be partly due to these times being usually periods of maximal activity in terms of grazing and social interactions. Sexual activity appears to be lower at night, particularly during the period from 11.00 pm to 3.00 am (Fig. 3.15).

Rams walk considerable distances when oestrous ewes are widely dispersed. In a study of rams in a 20 ha paddock, the distance covered by

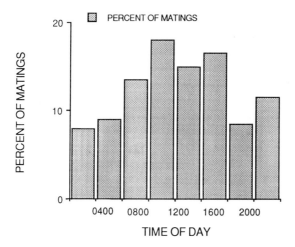

Fig. 3.15. Diurnal distribution of mating behaviour by penned rams (data from Hulet *et al.*, 1962).

both Merino and Poll Dorset rams was positively correlated (r=0.54 to 0.92) with the number of ewes mated per day (Raadsma and Edey, 1985). Rams travelled between 2.8 and 13.3 km per day; two to five times the distance travelled by ewes in the same flocks.

The number of coitions per ewe declines where rams are confronted with increasing numbers of oestrous ewes. Courtship behaviour is also curtailed where there are large numbers of ewes in oestrus in the first cycle of joining. Between-ram variation in the number of services per day is large. For example Merino rams ranged between 8 and 38 services per 13 hours in single-sire joining groups (Mattner *et al.*, 1967).

Other factors can influence the mating behaviour of rams. Sexual activity is depressed by high ambient temperatures, with many rams diverting from their sexual interests to shade-seeking behaviours. Consequently the number of ewes mated during a day declines and the probability of undetected oestrus increases. Inadequate nutrition can also influence sexual behaviour of rams. Penned rams appear to retain libido with liveweight losses of up to 15%. Weight losses beyond this appear to cause a dramatic decline in libido and in semen quality. Most field studies report ram weight losses of 10 kg or more through the first cycle of mating but with no adverse effects on libido. The rams normally regain weight in later cycles as mating and searching activity levels decline (Raadsma and Edey, 1985).

Age effects

The majority of mating management practices throughout the world recognize the need for particular care in the management of young and virgin animals. This results from the observation that conception rates in young ewes are normally lower than the rest of the flock and that young rams may not be able to compete for ewes with older rams.

Ewes

'Shy' breeding of maiden ewes has been reported repeatedly since the 1950s with young ewes showing an incomplete repertoire of sexual behaviour. They do not compete with older ewes for access to rams. Only experienced ewes seek out rams and compete for their attention. Young ewes also appear to disassociate themselves from rams after each courtship sequence and consequently the probability of the ram locating this ewe again is reduced. The chance of remating is low given the evidence that two-tooth ewes show irregular cycles with shorter oestrus of approximately 12 hours compared with 21 hours for mature ewes.

Consequently the following practices should be considered for mating maiden ewes:

1. use active, dextrous rams with persistent courtship activity;
2. mate maiden and adult ewes separately;
3. select paddocks which will ensure a high ram–ewe contact;
4. increase the ram:ewe ratio if paddocks are large.

These practices have been tested in a number of studies and it seems that paddock size is of little consequence at ram:ewe ratios of greater than 3:100 (Allison and Davis, 1976a). However, at lower ratios the number of matings per ewe declines markedly in the larger paddocks particularly for maiden ewes (Fig. 3.16).

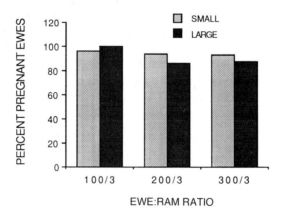

Fig. 3.16. The percentage of maiden ewes pregnant as influenced by ewe:ram ratio and paddock size (data from Allison and Davis, 1976b).

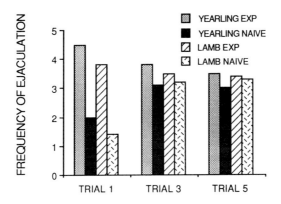

Fig. 3.17. The frequency of ejaculations by naive and experienced ram lambs and yearlings in a series of five 30-minute tests (data from Price *et al.*, 1991a).

Rams

The use of ram lambs in breeding programmes is limited because of the perception that young animals are relatively less competent. It has been found that eight-month-old rams which are sexually inept at initial exposure to oestrous ewes quickly attain a sexual performance similar to but slightly less dextrous than that of older rams (Fig. 3.17).

Since young rams are dominated by older animals in the field it is recommended that in multiple-sire matings the mixing of young and mature rams be avoided particularly where paddocks are small. In a comparison of young and mature Poll Dorset rams in multiple- or single-size joining groups, it was found that there was a high incidence of ineffective mating activities by young rams in the multi-sire groups when presented with 12 ewes in oestrus. This is shown in the mount to service ratios in Table 3.4.

Flock fertility may be increased substantially by decreasing the number of ewes mated per ram. Ram:ewe ratios of between 1:25 and 1:50 are necessary for young rams during their initial mating period if flock fertility is to be maintained at a similar level to that attained by mature rams.

In wild sheep, only older males attain access to ewes during the rut. In a

Table 3.4. Mount to service ratio for young and mature Dorset Horn rams.

Ram age	Young	Mature
Single-sire group	18:1	3.2:1
Multi-sire group	40:1	5.5:1

study of a Bighorn population where older males had been removed by hunting, Shackleton (1991) evaluated the aggressive and sexual behaviour of young males and found that males younger than 2.5 years exhibited normal courtship behaviours when the absence of older rams (> 7.5 years) allowed them to express these behaviours.

The influence of dominance on mating success

There is disagreement in the literature on the effect of dominance and competition on mating success in rams. One experiment has reported a relationship between dominance and mating priority (Lindsay and Ellsmore, 1968). This was most apparent at high ram densities or in confined areas where aggressive animals had greater access to oestrous ewes. The effect has been well illustrated in a study where ram numbers were small and the conception rate in a flock of ewes, with the dominant ram infertile, was reduced from 92% to 54% (Fowler and Jenkins, 1976). The tendency for ewes to be mated by more than one ram reduced the impact of dominance on conception rate when more than two rams were present with only a 12% reduction in the number of ewes pregnant. Such dominance effects can be mediated not only by the physical presence of the dominant ram but also by an 'audience' effect. Lindsay *et al.* (1976) have shown a marked suppression in the number of mounts of subordinate rams which were being observed by dominant rams (Fig. 3.18).

In larger flocks and in big areas there is no evidence of adverse effects of dominance on mating behaviour. The proportion of successful mounts may even increase where competition exists. In such competitive situations dominant rams have access to 'preferred ewes'. If subordinate rams are mated individually they will select 'preferred' oestrous ewes which will not

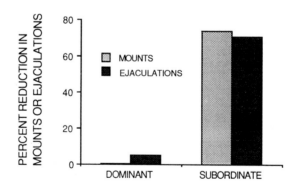

Fig. 3.18. The percentage reduction in mounting and ejaculation frequency of dominant and subordinate rams associated with the presence of 'audience' rams (data from Lindsay *et al.*, 1976).

be available to them if dominant rams are present (Synnott and Fulkerson, 1984).

The size and topography of the paddocks or fields used for joining are likely to affect mating success. In extensive grazing conditions these variables are linked with many others including feed availability and flock dispersion. This may be why it is difficult to find consensus on the effect of paddock size and ram ratio on mating success. The dispersal of ewes into sub-groups in response to low feed availability or heterogeneous environments may require an increase in the ram percentage required to retain ewe–ram contact.

Once dominance structures are established, the rams can search for oestrous ewes without disruption and the high-energy demands of further conflict. However, disruption of the mating behaviour of dominant rams can occur in situations where age structures favour younger males. In more confined conditions, dominant rams can suppress the mating activity of subordinates. In these situations the mating activity of the subordinate is constantly interrupted by the dominant ram.

Practically, dominance is of little consequence to mating management except in situations where the dominant ram is sterile for some reason. Acute disease with high temperatures can cause temporary infertility lasting for two to six weeks which results in a large proportion of the ewes returning to service. Dominance may be a problem if ewe numbers are small and an even distribution of sire matings is required. Mating management of multi-sire groups must therefore involve consideration of the area available per ram and the ram:ewe ratio in the flock as well as the topography of the mating paddock.

Tests of Serving Capacity and Libido

The practical implications of variability in ram libido are those associated with the prediction of the rams most likely to sustain mating performance in the field. Early studies characterized 'worker' and 'non-worker' rams as a first step in the prediction of field performance. A series of pen-testing procedures have been developed over the years in an attempt to predict mating performance. These include:

1. reaction-time tests – the time taken to first ejaculation;
2. exhaustion tests – maximum number of services possible per ewe;
3. mating dexterity tests – ratio of mounts to services;
4. serving capacity tests – services achieved in a set time.

All of these tests have relevance in particular situations, the first two being of importance for the assessment of rams for semen collection or pen-mating situations and the latter more relevant to field mating.

Considerable effort has been spent on examining the use of serving capacity tests as a selection criterion for rams and also as a 'screening' technique to identify poor or non-workers. To date the heritability of such tests in young rams has been shown to be low or close to zero, at least for Merino rams (Kilgour, 1985a). It seems that testing is most useful for screening against non-workers but not as a criterion for selection.

Serving capacity tests have been conducted using open or enclosed pens of various sizes, with a number of oestrous ewes and rams and tests over varying lengths of time. The most common test appears to be the one which has one ram with two or three oestrous ewes in an enclosed pen approximately 8 m × 5 m, for 20 to 60 minutes. An individual ram is introduced to the pen and the number of services in one hour is recorded. The refinement of testing procedures by the use of one-hour tests appears to have improved repeatability of such tests to a point where they can be used for 'on-farm' screening of rams.

Relationship between pen and field tests

The relationship of pen-test information to field-mating performance has shown varying results. Most of these tests have been done by identification of the 'best' and 'worst' rams in pen tests: these rams were subsequently used in field matings. The total number of services in 3 × 20 minute tests of Merino rams was highly correlated with the number of services observed in field matings in one trial (Mattner *et al.*, 1971; Fig. 3.19) but in other studies no relationship between pen-test performance and field-mating success has been found (Kelly *et al.*, 1975). These differences may be due to a variety of factors including the possibility that the tests did not differ-

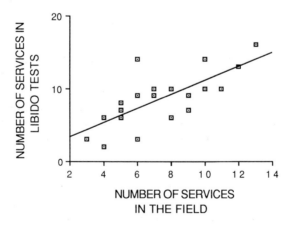

Fig. 3.19. The relationship between the number of ejaculations in libido tests and in a three-hour period on the first day of field mating (data from Mattner *et al.*, 1971).

entiate between rams or that mating pressure in the field was not great enough to allow expression of the differences observed in pens. It is now recognized that many non-workers will become workers after a number of exposures to oestrous ewes and the repeatability of test rankings increases as rams become older or have greater exposure to oestrous females. Kilgour *et al.* (1985) reported that in a group of non-working Border Leicester rams, 87% began to mount and serve on a second exposure to oestrous ewes.

Pen-mating Systems

Pen mating is used in stud and research situations to ensure the pedigree of lambs conceived and to maximize the use of individual rams. It is essential for the animals to be adapted to their surroundings. This may take up to two weeks. Pens should be visually separate from one another to prevent dominance effects of other rams although this may not be absolutely necessary as 'trained rams' will work well even in relatively confined areas and within sight of other rams.

Unlike bulls and bucks, young rams are only slightly sexually stimulated by the sight of other rams mounting (Price *et al.*, 1991b) and therefore the common practice of more than one male used to improve sexual activity of bulls is inappropriate in this species.

There are few breed differences in mating behaviours which will influence conception rates achieved in pen-mating situations. Winfield and Kilgour (1977) identified only small differences in the number of services and in reaction times of rams of eight different breeds used in a pen-mating programme.

Semen collection

Semen collection procedures require an awareness of the mating behaviours of the ram since a number of problems can arise with routine collection procedures. These include reduced libido through habituation and stereotypes resulting from a restrictive and 'dull' pen environment. Knowledge of the behavioural cues which stimulate rams will allow reinforcement of arousal responses when training rams to serve into an artificial vagina (AV).

Training and overcoming inhibitions

Routine semen collection requires that rams be conditioned to mount and ejaculate in response to an immobile or dummy ewe. A necessary part of the training process is reinforcement of the normal behavioural sequence which is facilitated by the presence of an experienced oestrous ewe.

Young rams can present problems as they tend to be slow in their initial responses; patience is required. The use of non-tethered ewes in a larger area allows young rams to establish courtship behaviours which lead to correct orientation and mounting. After young rams have established normal patterns they can be introduced to the immobilized teaser ewe and allowed to mount in the absence of the handler. When the rams have become adept in mating in this situation they must then be habituated to the presence of handlers. Only then should semen collection be attempted.

There can be considerable variation between rams in their response to training for semen collection. Patience and a 'slow and steady' approach is necessary to evoke the sexual responses required. Major breed differences in ease of training are not evident and the greatest proportion of variability appears to be related to age and fear responses to the handler.

Collection

Once the ram has overcome the initial fear of the handler and surroundings and has learned to mount, collection procedures can be initiated. The four stages of activity leading to ejaculation need to be recognized so that semen can be collected with a minimum of fuss and maximum efficiency.

1. Courtship behaviours precede mounting and seem to be a major source of excitation for both ram and ewe. These behaviours were described earlier in Table 3.2. Rams vary considerably in the time taken in courtship behaviours and in some trained, older rams these behaviours can be almost non-existent. If rams do not show normal excitability the introduction of a new oestrous ewe may initiate activity.
2. Once the ram is excited he will attempt to mount. This act is usually preceded by erection of the penis. The artificial vagina should not be used until the ram is mounting. It is important that the ram is not exposed to negative stimuli such as pain during deflection of the penis into the artificial vagina.
3. Ejaculation will usually follow rapidly after the penis is guided into the artificial vagina. At this stage rough handling, an inadequately prepared artificial vagina or any major disturbance of the ejaculatory reflex are negative stimuli which are likely to inhibit the ram in subsequent collections.
4. The post-copulatory period should not be interrupted as this is a period of positive reinforcement. The ram and ewe should remain together for a few minutes before separation.

Rams can ejaculate four or five times before the latency or refractory period exceeds 20 minutes. The presence of a new ewe may restore activity for another one or two ejaculations. Between-ram variation may be large and thus high numbers of semen collections per day are not usually employed.

Conclusions

Our knowledge of the behaviour characteristics of sheep associated with mating is relatively complete. We not only have an understanding of the behavioural patterns but also of the basic psychological processes which control them.

In modern sheep husbandry systems the introduction of technologies to synchronize the oestrous cycle of the ewe, the desire to reduce ram:ewe ratios and the development of semen collection and pedigree mating practices have focused attention on the reproductive capacities of the ram. Most of the recent studies of sheep behaviour have been orientated to the active role of the male rather than the more passive behavioural role of the ewe.

Even though the normal mating behaviour patterns, as exhibited by feral and wild sheep, are disrupted by many husbandry initiatives, innate behavioural drives of the sheep seem to override this disruption. Consequently the *Ovis* species are able to adapt mating behaviour patterns to a wide diversity of situations and environments.

Further Reading

Banks, E. (1964) Some aspects of sexual behaviour in domestic sheep, *Ovis aries. Behaviour* 23, 249–79.

Fowler, D.G. (1984) Reproductive behaviour of rams. In: Lindsay, D.R. and Pearce, D.T. (eds), *Reproduction in Sheep.* Australian Academy of Science, Canberra, pp. 39–46.

Grubb, P. (1974b) Mating activity and the social significance of rams in a feral sheep community. In: Geist, V. and Walther, F. (eds), *The Behaviour of Ungulates and Its Relation to Management.* ICUN Publications, Morges, Switzerland, pp. 457–76.

Kilgour, R. (1985b) The behavioural background to reproduction. In: Fraser, A.F. (ed.), *Ethology of Farm Animals*, Elsevier, Oxford, pp. 279–88.

Lindsay, D.R. (1985) Reproductive anomalies. In: Fraser, A.F. (ed.), *Ethology of Farm Animals.* Elsevier, Oxford, pp. 423–9.

4

The Pregnant, Parturient and Lactating Ewe

A time span of four weeks from the birth of the lamb is critical to the survival of lambs and to the economic success of domestic sheep producers. During this period some ewes, and about 15% of lambs will die. Mortality may be less for sheep such as the Mountain Bighorn, Soay and Mouflon, which may be able to optimize their responses to the environment and their lambs without impediment from people. The maternal behaviours which are involved in normal ewe–lamb associations from birth to weaning and the modification of behaviour with breeds and the environment will be discussed in this chapter. Any description of the behaviours associated with parturition and bonding presents problems in generalizing from studies made in a range of different climates, topography and nutrition as well as different group and litter sizes. The reason is that conflicting needs for food, shelter, water and conspecifics may be experienced in different combinations by ewes in the many environments in which sheep are found. Furthermore, management may constrain the full expression of particular behaviours. Another limiting factor in behaviourial studies on ewes is the extreme difficulty of observing parturient wild Asiatic or American sheep.

Pre-parturient Behaviour

The gestation length in sheep varies from 145 to 155 days in domestic breeds and can range to up to 170 days in some wild breeds. Pregnancy is shortened by one to two days if the ewe is carrying more than one lamb.

The lambing behaviour of seven breeds of domestic sheep was observed by Arnold and Morgan (1975) and although there was considerable variation in the length of pre-parturient behaviours, the same behaviours were seen regardless of breed or age of the ewe. Pre-parturient behaviour of ewes includes alternately standing and lying, walking in circles, sudden movement for about 10 m, licking the lips or making a tongue movement into the air, licking the ground where amniotic fluid has been spilt, pawing the ground and making distress calls.

An initial sign of imminent parturition is restlessness. This is shown by two-thirds of ewes and generally occurs within three hours of parturition. The period varies between individuals and lasts from 10 minutes to 11 hours.

Another pre-parturient behaviour is the interest shown in the newborn lambs and the amniotic fluid of other ewes. This behaviour, induced by elevated levels of oestrogen in the ewe within 12 hours of parturition, can also include attempts to foster another ewe's lamb. If the ewe is successful in stealing the lamb then the latter may die since the ewe often loses interest in the other lamb when its own parturition is imminent. Alternatively, she may neglect her own lambs. This interference with alien lambs is more commonly seen when there is a high density of parturient ewes either because of high stocking rates or because of deliberately synchronized lambing.

Isolation

Isolation of the ewe from the flock is often another sign of parturition and is important for the subsequent rapid development of ewe–lamb recognition without other parturient ewes attempting to disrupt the association developing between the ewe and its lamb. This trait appears to increase the chances of lamb survival.

Isolation of the parturient ewe was clearly described for Oxford and Cheviot ewes by Fraser in 1926. Rocky Mountain Bighorn ewes have been observed to move away from the flock up to two weeks before parturition.

Arnold and Dudzinski (1978) point out that much of the described variation in isolation behaviour between breeds may be related to the definition of isolation by various authors. Unless observations are continuous and deliberately measure whether the ewe moves away from the flock rather than that she is left behind as the flock moves away, it is difficult to say that a ewe isolates herself from the flock.

When this behaviour is assessed by measuring the location of the flock and the ewe at the same time, ewes on the point of parturition do tend to seek isolation from the rest of the flock. Soay, Lacaune, Scottish Blackface and Mouflon ewes all isolate themselves from the flock. Isolation at lambing was also observed in an experiment in which Suffolk and Targhee ewes were housed. Some sheep lambed in a large shed (control) while others lambed in a similar shed, but could choose to use small cubicles which were located around the periphery of the shed or lamb in the body of

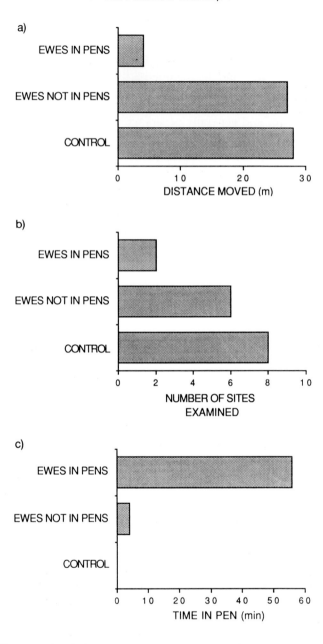

Fig. 4.1. Pre-parturient behaviour of ewes: (a) distance moved; (b) number of sites examined; and (c) time in pen (data from Gonyou and Stookey, 1985).

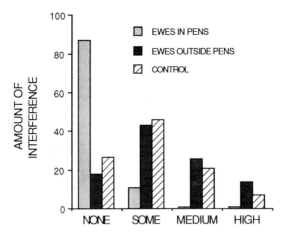

Fig. 4.2. The number of times other ewes interfere with the parturient ewe which is inside the pen, outside the pen, or has no pen available (data from Gonyou and Stookey, 1985).

the shed where they intermingled with other ewes. The pre-parturient behaviours of the control group and those ewes which did not use the cubicles were similar. Half the ewes with access to the cubicles lambed in them. There was less movement in this group compared with the other two groups (Fig. 4.1). Over the four hours after parturition, ewes choosing the cubicles stayed closer to their lambs and experienced far less interference from other ewes (Fig. 4.2). No sheep lambing in the cubicles were separated from their lambs or had them stolen while more than 5% of multiples were stolen in each of the other two groups and may well have died if it had not been for the strict management imposed on the experiment. If sheep are housed there are considerable benefits in providing individual pens for lambing sheep.

Considerable effort has been spent determining if fine-wool Merino sheep lamb in isolation. In one experiment (Stevens *et al.*, 1981) only 10% isolated themselves. However, in two other experiments (Alexander *et al.*, 1990) about half the sheep isolated themselves from the flock, a figure more in keeping with that observed in other strains of Merino. In the latter two experiments, the paddocks were not flat, featureless areas with a fence around them, as in the first experiment, but contained topographical features of trees, hills and rocks. The importance of these features on parturient behaviour has not been assessed.

Birth site

In flat, featureless country, the location of the birth site is distributed over the whole area but where there is even a slight elevation lambing occurs in the more 'elevated' area. When sheep lamb in hilly country their lambing sites are near fence lines and along the ridges. Within these 'preferred' areas there will be a close aggregation of birth sites probably because fresh fetal fluids from previous births attract parturient ewes. Also these birth sites are re-used in subsequent years. In small experimental paddocks ewes predominantly lamb within 6 m of fence lines (Fig. 4.3).

Parturient Behaviour

Although a great range of behaviours are shown by the ewe around parturition, there is no way that an experienced observer can accurately predict the time of parturition. The types of parturient behaviour which include pawing at the ground, alternate standing and lying, walking in circles and vocalizing are consistent between breeds. Labour is generally completed within an hour. The time taken both from the commencement of intense straining and the appearance of the feet at the vulva until birth is shown in Fig. 4.4 for ewes with different litter sizes. Those ewes with more than one fetus have subsequent lambs born more rapidly than the first with the time interval between births being about 20 minutes. Some higher litter size lambs are born rapidly and with very little apparent effort while the mother is licking another lamb. The time from the initial straining to completion of

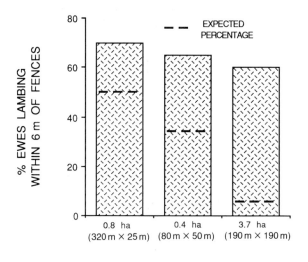

Fig. 4.3. The percentage of ewes lambing within 6 m of the fenceline in paddocks of different area.

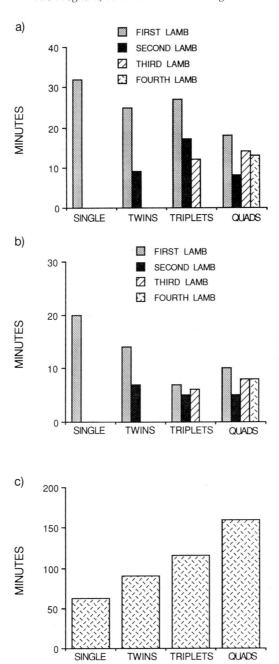

Fig. 4.4. The time taken (minutes) for ewes with singles or multiples to deliver their lambs from the start of (a) intense straining and (b) appearance of feet at the vulva. The duration of the whole birth process is shown in (c) (data from Owens *et al.*, 1985).

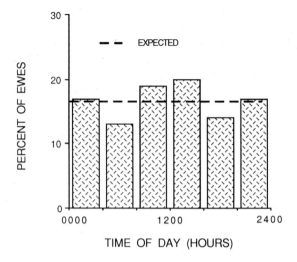

Fig. 4.5. The distribution of Merino births over 24 hours together with the line showing the expected distribution (data from Owens *et al.*, 1980).

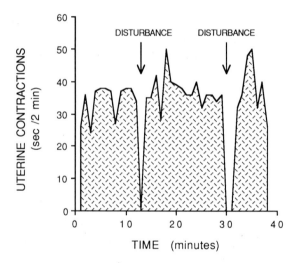

Fig. 4.6. The effect of disturbance on uterine activity (adapted from Naaktgeboren, 1979).

the birth process varies from about 70 minutes with singles and from 150 minutes with quadruplets (Fig. 4.4).

Primiparous ewes can have slightly longer labour than multiparous ewes. A likely reason is the smaller size of the birth canal related to the size of the fetus. Primiparous ewes may also startle and run nearing the end of the birth process. This is particularly so if the ewe is standing and the almost-born lamb hits the ewe's back legs. Fortunately it is rare for the ewe to desert the lamb when this occurs.

A long labour can be associated with oversize fetus, malpresentation of the fetus or undernutrition during pregnancy. In the latter situation the ewe spends long periods without straining to expel the lamb. Whatever the reason for the long labour, the frequent result is subsequent disinterest in and desertion of the lamb.

There have been reports of a higher probability of parturition at particular times of the day. Hampshire and Dorset Horn are more likely to have lambs in the morning. When the distribution of time of lambing was examined over many births in Merinos and Finnsheep, there was very little periodicity whether the ewes were left alone or were attended over the whole 24 hours (Fig. 4.5). The disturbance of the parturient ewe by an observer resulted in uterine inertia and inhibition of labour and it took some time for the uterus to resume contractions (Fig. 4.6). When research workers do not maintain a 24-hour vigil on pregnant ewes, there will be more lambs born when they are absent than when they are present.

Shelter-seeking behaviour

There is a heavy mortality in the neonate if the weather is windy and wet. This mortality is illustrated in Fig. 4.7 which shows the percentage of Corriedale and Merino lambs which die if they are born with no shelter

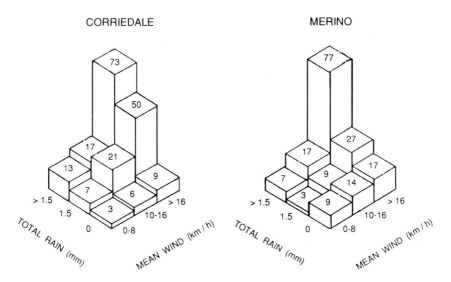

Fig. 4.7. The association of post-partum lamb mortality over 24 hours of Merino and Corriedale with mean wind speed and total rain in the first six hours of life (reproduced with permission from Obst and Ellis, 1977).

Fig. 4.8. The use of shelter made by recently shorn lambing ewes.

Fig. 4.9. Five-year-old shelter belts used for lambing.

available. Seventy-five per cent of lambs die if they experience more than 1.5 mm of rain and winds greater than 20 km/h during the first six hours after birth.

If there is no rain, wind speeds less than 40 km/h do not cause pregnant Corriedale, Merino or Scottish Blackface to seek shelter, but Welsh Mountain ewes are said to seek shelter at lower wind speeds. Wild and feral sheep tend to lamb in shelter. Rocky Mountain Bighorn ewes select broken rugged cliffs as lambing sites where they will inevitably be better protected than in the surrounding high plains (Geist, 1971). Soay sheep select protected areas of man-made drystone to lamb during inclement weather (Grubb and Jewel, 1966). Shelter-seeking is directly correlated with nutrition so that the undernourished sheep are much more likely to shelter from wind and rain than sheep in good condition. When Merino sheep are shorn within six weeks of lambing they will move quickly to sheltered areas during light rain and windy conditions, but if rainfall intensity increases ewes may be driven out of poor shelter to lamb in an exposed area near a fenceline. If shelterbelts are located 20 m apart, shorn sheep use them except when grazing and thus a high percentage will lamb there simply because the majority of the 24 hours is spent near shelter (Figs 4.8 and 4.9) (Lynch and Alexander, 1977). One result of this is a higher survival rate of lambs born in these areas.

Description and Control of Maternal Behaviour

Post-parturient behaviour

Most ewes stand and commence licking their lambs vigorously within two minutes of the birth. This rapid licking is driven by an intense attraction to birth fluids (Levy and Poindron, 1987). It dries the lamb and results in ingestion of much of the fluid and membranes. The licking and the olfactory stimuli presented to the ewe result in maternal recognition of the lamb by the time the lamb sucks. The attentive licks are accompanied by a change in the ewe's vocalizations to a deep soft bleat or rumble. In experimental studies of maternal behaviour which will be discussed later, licking, bleating and acceptance of the lamb have been used as criteria of a positive response. Absence of maternal behaviour is demonstrated by rejection which may, at times, include violent butting of the lamb.

The first site the ewe licks is almost always the head and neck of the lamb. Vince *et al.* (1985) examined the number of 10-second periods in which the newly parturient ewe licked: (i) the head and neck; (ii) the forelegs and flank; and (iii) the hindquarters of its single lamb during the first hour of life. During sequential 15-minute intervals there was a marked drop in the amount of licking to the head and trunk of the lamb and a

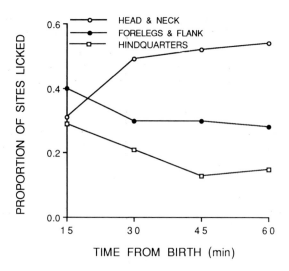

Fig. 4.10. Proportion of grooming directed at each of three sites on the lamb during the first hour of life.

marked increase in licking of the hindquarters (Fig. 4.10). There is no suggestion that sheep lick from the head progressively to the tail since the amount of attention given the posterior of the lamb is always high. In the first hour there is a higher overall licking activity in Border Leicester × Merino ewes compared with Merino ewes.

Initiation of maternal behaviour

The urgency for a strong association to form between the ewe and its newborn lamb arises because ewes of most domestic breeds rejoin the flock within 12 hours of parturition. The survival of the lambs is dependent upon the ewe knowing her lambs and ensuring that they are close to her at all times and that no separation occurs. In the short period before and after parturition almost all parturient ewes show maternal behaviour towards any newly born lamb. This behaviour is sudden in its onset and wanes quite rapidly if the young lamb is removed. Its continuation is stimulated by the lamb's presence.

The speed with which the attachment between the ewe and her lamb occurs has been stated as 20 to 30 minutes after the lamb's birth. The fact that the lamb is allowed to suckle seems a definite indication that the ewe recognizes and accepts her lamb. There is evidence that the first hour post-partum is important in the establishment of a selective recognition between the ewe and its own lamb. If lambs are present with the mother but contained within a double wire cage, the ewe establishes a very selective association, but if the lamb is withheld and placed in the cage one hour after birth the ewe becomes less selective and accepts any lamb as its own (Alexander *et al.*, 1986). Presumably lamb odour is involved.

Behaviour which prevents the lamb sucking, such as circling behaviour, is considered abnormal. Perhaps this behaviour is related to the ewe failing to recognize its lamb as it occurs very early in the association between dam and lamb.

If ewes are separated from their lambs for the first 24 hours of life, half may not accept their own lamb. If, however, lambs have been left with their mothers for a short period at birth and then removed for many hours, lambs will readily be accepted upon return and allowed to suck. Levy *et al.* (1991) have shown that when a lamb was removed four hours after birth, the mother would readily re-accept the lamb providing it was returned within 24 hours. After this time, maternal behaviour and selectivity for her own lamb waned rapidly.

Control of Maternal Behaviour

Sensitive period

A distinction has been made between the initial phase of maternal behaviour which corresponds with the onset of parturition and wanes rapidly afterwards and the maintenance phase which continues, in the presence of the ewe's offspring, until weaning. The first phase has been called the 'critical' or 'sensitive' period and is generally under hormonal control while the later phase is considered to be largely non-hormonal in nature but more psychosensory. Excellent reviews of the factors initiating and maintaining the sensitive period of maternal behaviour in sheep have been given by Poindron and Le Neindre (1980) and Poindron and Levy (1990).

Hormones

Studies of the factors inducing maternal behaviour in the ewe have revealed the importance of fetal fluids, olfaction and the birth process which cause changes in hormone levels. These, together with experience of parturition, are integrated and eventually processed in the brain.

Maternal behaviour and the role of hormones have been intensively studied in sheep over the last 15 years using the criteria of the ewe licking the lamb, the emitting of low-frequency bleats and allowing the lamb to seek the udder and suck. The relation of maternal behaviour to levels of blood oestrogen and progesterone at the various stages of the reproductive cycle is shown in Fig. 4.11. It is possible to induce maternal behaviour in more than 50% of ovariectomized non-pregnant ewes which are injected with progesterone and oestrogen.

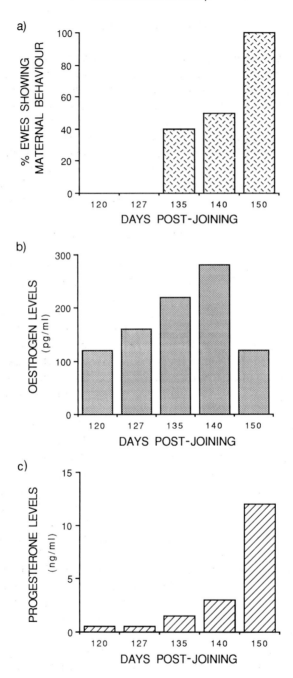

Fig. 4.11. Maternal behavioural responses of ewes towards newborn lambs at various stages of the reproductive cycle (a), and in relation to blood levels of oestrogen (b) and progesterone (c) (data from Poindron and Le Neindre, 1980).

Vaginal stimulation

From time immemorial shepherds have induced acceptance of alien lambs by placing a hand into the cervix of a recently parturient ewe. The physiological and behavioural consequences of dilation of the cervix and vagina which occurs during birth have been reviewed by Keverne and Kendrick (1990). Vaginal stimulation (VS) induced maternal behaviour in a group of non-pregnant ewes primed with oestrogen and progesterone (Keverne *et al.,* 1983). Figure 4.12 shows the change from behaviours associated with rejection of lambs to strong maternal behaviour in ewes after stimulation of the vagina. Vaginal stimulation also results in the adoption of an alien lamb after the ewe has formed a strong selective association with its own lamb.

Central processing of hormones, olfaction and vaginal stimulation

Two major reviews, Poindron and Levy (1990) and Keverne and Kendrick (1990) have highlighted the research which has extended the initial descriptive studies on the behaviour of the ewe at parturition to an explanation of the role of hormones and birth fluids on lamb recognition and rejection. They have also integrated the neuro-hormonal mechanisms within the brain, the physical stimulation of the birth process and the olfactory system to explain many aspects of maternal behaviour. Poindron *et al.* (1988) correctly state that they have opened up research areas of 'functional anatomy of the sheep brain, especially oestrogen-sensitive structures, oxytocin-containing sites and pathways, aminergic pathways and their interactions'.

Oxytocin is liberated into the peripheral circulation during the last stages of labour and by VS. It is also liberated within the brain under the same circumstances. Since oxytocin will not pass the blood-brain barrier, there must be two separate sites for production of the hormone.

Recent work has shown that genital and olfactory stimulation of the parturient ewe interact within the brain under the influences of oestradiol and oxytocin to cause observable maternal behaviour. In this regard, oxytocin release from the posterior pituitary is associated with VS in oestrogen-primed ewes. In addition, an intracerebroventricular (ICV) injection of oxytocin has been shown to induce maternal behaviour. The importance of ICV oxytocin to maternal behaviour of the rat has been confirmed by giving anti-oxytocin antiserum in ICV-primed animals. This antiserum significantly diminished maternal behaviour which was reversed by the addition of more oxytocin.

Vaginal stimulation may also induce maternal behaviour via another mechanism related to the olfactory bulb and thus initiate specific recognition of the ewe's lamb. Activation of noradrenaline fibre projections to the olfactory bulb depends on VS. Further, the noradrenaline input within the olfactory bulb is important for memory and hence recognition of the

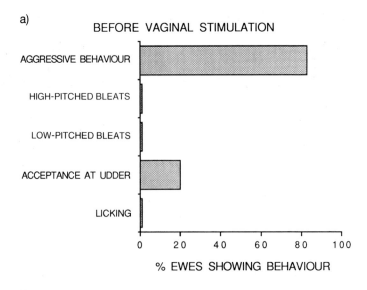

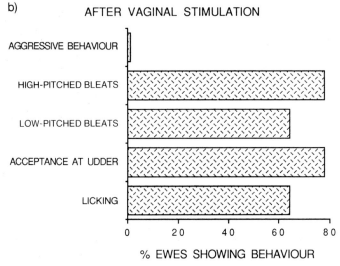

Fig. 4.12. Maternal behaviour of non-pregnant ewes before (a) and after (b) vaginal stimulation (data from Keverne *et al.*, 1983).

newborn via the amniotic fluid specific to the newborn lamb. Creating a lesion between the noradrenergic projections and the olfactory bulb prevents the parturient ewe from forming a selective bond with its lamb and results in her accepting alien lambs.

A comprehensive treatment of neural and neurochemical mechanisms and their interaction in the onset and intensity of maternal behaviour has

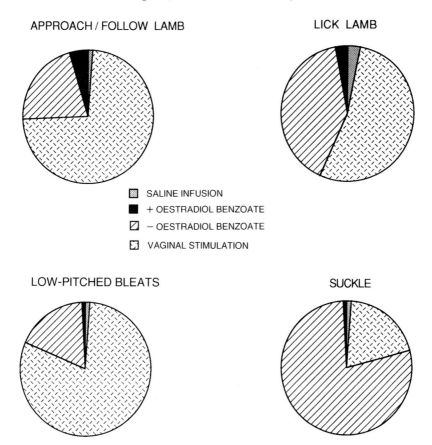

Fig. 4.13. Mean frequency of acceptance behaviours following intracerebroventricular oxytocin infusion, with or without oestradiol benzoate and/or saline. These are compared with vaginocervical stimulation (data from Keverne and Kendrick, 1990).

been written by Keverne and Kendrick (1990). Injections of oxytocin (OT) and oestradiol benzoate (OB) have resulted in non-pregnant ewes being attracted to a newborn lamb. One minute after the injection of OT, the ewe showed many aspects of maternal behaviour as described earlier and few aspects of lamb rejection (Fig. 4.13).

Stimuli provided by the lamb

As the ewe can be considered as a composite stimulus to its newborn lamb, so too the lamb appears to provide many stimuli for the ewe. Although amniotic fluid is most attractive to the ewe, anosmic ewes do accept their lambs. Lamb bleats have been observed to attract parturient ewes thus

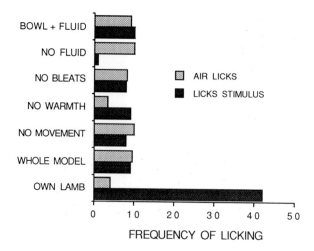

Fig. 4.14. Frequency which ewes nosed or licked the stimulus or licked the ground or made licking movements in the air.

demonstrating a role for auditory stimuli. Merino ewes soon abandon still-born lambs, so warmth and movement may also play a part in acceptance.

In research by Vince *et al.* (1985) the sensory stimuli provided by the neonate was broken into its major components and the parturient ewe's response to these stimuli was examined. Lambs were removed from ewes at birth and groups of ewes were subsequently presented with (i) own lamb, (ii) a model lamb which was warm, moved, bleated and was covered by amniotic fluid, (iii) a model lamb with one of the four parameters of (ii) removed or (iv) a white bowl of amniotic fluid. The dam's approach, licking and nosing the lamb or the model was recorded every five seconds for five minutes. The results were quite unambiguous. Although the amniotic fluid alone presented an important stimulus, the ewes were far more attentive to their own lamb than to any other group of stimuli (Fig. 4.14).

Effect of parity

Maiden ewes are more likely to desert their lambs than those ewes which have experienced a previous parturition. Furthermore, there are more disturbances to the normal behaviourial sequence and more lamb deaths in primiparous and thus inexperienced ewes.

The importance of previous experience of lambing has been highlighted by some experiments on the parturient ewe and its neonate. The absence of amniotic fluid on the newborn or the lack of vaginal stimulation during parturition caused by an epidural anaesthetic (Krehbiel *et al.* 1987) and

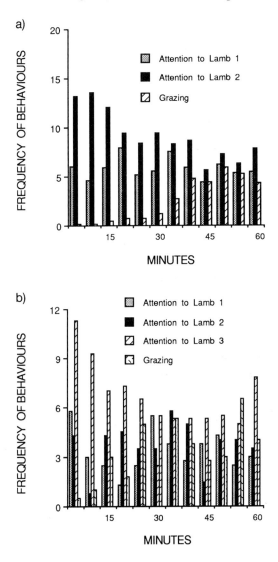

Fig. 4.15. The attention the ewe gave in the hour after the last lamb was born to (a) the first or second twin or (b) the first, second or third triplet. The alternative behaviour was grazing.

caesarian section (Alexander *et al.*, 1988) created more disturbances to normal maternal behaviour in the primiparous, compared with multiparous ewes. Presumably the complex series of stimuli are all needed in the primiparous ewe to ensure the initiation of normal maternal behaviour which, in itself, is innate. With experience, the sequence of stimuli can have parts omitted and normal maternal behaviour will still be expressed. Previous experience of twins is also important in the ability of ewes to switch to grooming a newborn second twin.

Litter size and maternal behaviour

The level of attention given to twins in the hour after the second twin was born and to triplets in the hour after the third was born is shown in Figs 4.15a and b. The last born twin or triplet received significantly more attention than other lambs for the first 30 minutes after birth. It then received as much attention as the first born twin or the first two triplets. Grazing increased significantly during the last half hour when the last lamb born was observed. During the parturient period, total care giving time to the second or third born twin or triplet is much less than for the single. This may well have implications for the formation of associations between lambs and dam (Chapter 2).

Time at birth site

Most breeds of sheep spend long periods of time at the birth site, but there are differences between breeds. Bighorn ewes and their lambs remain for five to seven days while Soay and Mouflon ewes stay for two to three days. In a flock of New Zealand Romney sheep it was shown that the primiparous ewes stayed on the birth site for about 11 hours and that multiparous ewes stayed seven hours (Alexander *et al.*, 1983). By contrast, Merinos and Border Leicester × Merinos remain only four hours or even less than this if there is insufficient food at the site. The survival of a Merino twin lamb is strongly related to the length of time ewes remain on the birth site (Fig. 4.16).

When ewes with twin lambs were either (i) penned on the birth site, (ii) moved about 20 m away, (iii) moved and later returned, or (iv) were left alone, there were significant effects on maternal behaviour which resulted in the permanent desertion and death of 3, 18, 23 and 12 percent, respectively, of the lambs (Putu *et al.*, 1988). The average time spent on the birth site by the control Merino ewes was four hours. Half the ewes which were removed came back and licked the birth site. The attraction of the parturient ewe to amniotic fluid may well be a factor in the ewe remaining on the birth site.

Abnormal maternal behaviour

Primiparous ewes are more likely to show behaviours which result in death to the lamb. Desertion of the lamb can occur with two days of parturition and may either be temporary or permanent. Some ewes in poor condition

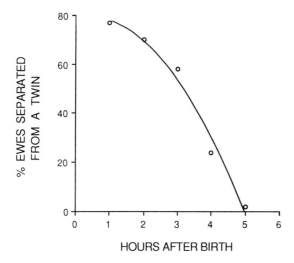

Fig. 4.16. The percentage of ewes which were permanently separated from a twin lamb in relation to the time the ewes spent on or within 20 m of the birth site (adapted from Alexander *et al.*, 1983).

or after a difficult birth walk away from their newborn lamb after an occasional lick or even without smelling the lamb. A ewe may also intersperse grazing with attention to the lamb. If the latter is relatively immobile the ewe gradually resumes full-time grazing and walks to join the flock (Fig. 4.17). Other ewes butt the newborn lamb vigorously if it moves even though they may groom it. Some ewes fail to stand still when the lamb attempts to suckle the udder. After some time, these behaviours usually disappear and the ewe allows the lamb to suck.

When ewes and lambs are penned individually an excellent mother/young association eventually develops. However, with sheep in paddocks or housed in large pens these abnormal behaviours can result in death of the lamb.

Arnold (1985) has classified as abnormal the failure of fine-wool Merino ewes to keep twins together. This is a behaviour which has not been reported from other strains of Merinos or other breeds of sheep. It occurs during the first 24 hours after birth and is exhibited when ewes which move from a resting place are accompanied by only one of a set of twins. It can be either twin which is left behind. In these experiments (Stevens *et al.*, 1982), hunger induced by lactation together with a low pasture availability may have caused the ewes to leave the birth site and graze, thus interfering with the maternal–young association. Other researchers (Putu *et al.*, 1988) studying the Merino have found that a low level of nutrition resulted in a

Fig. 4.17. The abandonment of a lamb.

higher proportion of permanent separations of at least one twin (19%) compared with ewes given a high level of nutrition (4%). A further problem for parturient Merino ewes may be that, in small flat paddocks with no trees, the drive for ewes to flock together becomes greater than their maternal drive. This causes ewes to leave the birth site before strong bonds have formed with both twins.

Cessation of sensitive period

There is some information about the length of time the sensitive period persists after the birth of lamb. It appears to depend on the timing of the lamb's removal from the ewe. Levy *et al.* (1991) have shown that removal four hours after parturition results in continued maternal behaviour for 24 hours before there is a decline in the number of ewes accepting lambs. If there is no contact between ewe and lamb after parturition, ewes reject their lambs by 12 hours. Since suckling and oxytocin production are well integrated and since the latter is clearly associated with maternal behaviour, it is quite possible that suckling the lamb will result in a waning of the sensitive period rather than its abrupt cessation. This waning could presumably be also related to the gradual reduction in oestradiol of the post-parturient ewe.

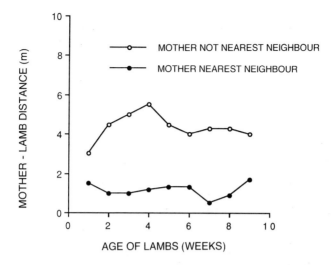

Fig. 4.18. Mean distance between mother and lamb with increasing lamb age.

Maintenance Phase of Maternal Behaviour

The behaviour of the ewe towards its lamb gradually changes with time. Ewes allow their lambs to suckle at any time during the first week of life but gradually restrict the frequency and duration by walking away from the lamb. During the first few days the ewe and lamb remain close together. When walking, the lamb is no more than one metre behind the ewe and normally close to the foreleg. When lying, the ewe and its lamb are together. Gradually the ewe leaves the lamb to graze in its vicinity and eventually may graze up to 50 m away. Over the first week, lambs congregate one or several times during the day and gambol prior to rejoining their mothers. Over the first month of life there is an increase in the ewe/lamb distance (Hinch *et al.*, 1987). This is followed by a decrease in the distance apart which is maintained until weaning. However, about 75% of the time, when the mother and her lamb are nearest neighbours, they are consistently less than two metres apart (Fig. 4.18).

By tethering lambs, first at 10 m from their mothers and then 30 m, it was shown that the ewe actively seeks its lamb for the first month of the lamb's life by coming up to it and bleating. Thereafter an increasing proportion stopped at half the distance between the release point for the ewe and the tethered lamb and bleated for the lamb to come before either turning to walk away or walk straight past their tethered lamb.

This behaviour represents an apparent reversal of ewe and lamb roles in the maintenance of their association. Since the ewes and their lambs were very close in the field for 75% of the time, it seems that the ewe maintains this close distance early in the lamb's life and thereafter the lamb plays an increasing part in maintaining the mother–lamb attachment. This changing pattern of mother–lamb behaviour has been reported for Lacune and Romney sheep, suggesting that the changing mother–lamb role in mainten-ance of ewe–lamb attachment is a normal behaviour of sheep, perhaps with a different time-scale of change between breeds.

Lamb recognition in the maintenance phase

The sense of olfaction in stimulation of maternal behaviour and in the immediate recognition of the ewe's own lamb has been previously described. The ewe quickly becomes adept in recognizing her lamb. In the first weeks of life the ewe rapidly sniffs the area around the lamb's tail as it moves in to suck the udder, although the ewe can recognize her lamb by smelling other areas of the body.

Vision and sound play an important role in the ewe recognizing her lamb from afar. The role of hearing and vocalizations, together with breed variations, has been described in Chapter 3. Research by Alexander and Shillito (1978) showed that changing the colour of lambs by using white, yellow, green, brown, red and black powder reduced the speed of approach of the mother and also introduced some hesitation or avoidance of the lambs whose ages ranged from 4 to 11 days. There is a clear ranking in the speed of acceptance from lighter to darker colours. All ewes eventually accepted their lambs within five minutes.

Of more importance was the identification of the area of the lamb that the ewe recognized. The blackened areas are shown in Fig. 4.19a. The speed of acceptance or rejection of the lamb which had various parts dyed black is shown in Fig. 4.19b. This figure clearly shows that the ewe was recognizing its lamb from visual clues around the head region. This recognition was confirmed by experiments on Jacob ewes who have black and white lambs. The white areas of the lamb's head were blackened and the behaviour of the ewe changed to one of hesitation and avoidance before accepting the lamb again.

The senses of smell, sight and hearing ensure that the ewe can seek out her lamb and allow it to suckle within the first three to four weeks of the lamb's life. After this time, the lamb must respond to its mother's bleats.

Weaning

The management of domestic sheep frequently involves the compulsory weaning of lambs at ages which have varied from four weeks old and older. The ruminant's digestive system has fully differentiated and development is almost complete by eight weeks of age (Lyford, 1988). Weaning by

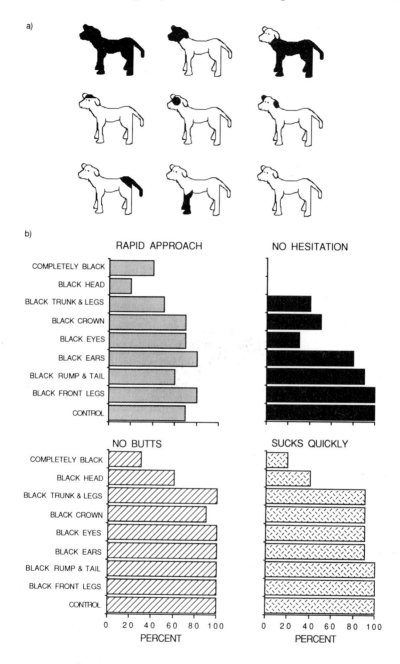

Fig. 4.19. (a) An illustration of the sites where lambs were blackened. (b) The percentage of ewes with a rapid approach, no hesitation, no butts, or suckles within 15 sec when confronted with their lambs which had a variable amount of black dye on them (data from Shillito-Walser and Alexander, 1980).

removing the ewe from its lamb has considerable consequences on the feeding (Chapter 1) and social (Chapter 2) behaviours of lambs. Many aspects of diet selection and, presumably, paddock geography are learned by lambs while being associated with their mothers. Weaning by separation may impose stress and have consequences for the poor development of immunity to parasites.

When mother and young have not been separated, a major determinant of the occurrence of weaning is the ewe's milk supply. The relation between milk supply and weaning of Dorset Horn lambs is shown in (Fig. 4.20). A major variation in the time of weaning will be related to volume of milk available (Arnold *et al.*, 1979). Ewes wean their lambs by preventing them sucking, a process which takes about seven days. After weaning, the ewe and lamb still associate closely and even if separated physically the lamb will return to the mother after two months (Hinch *et al.*, 1987).

Mountain Bighorn sheep wean their lambs around 120 days or earlier depending on the environment which influences milk production through herbage availability. Those sheep on better quality diets produce more milk which results in lambs being weaned at a later age (Berger, 1979). The interaction between weaning and subsequent oestrus which is often synchronized in wild sheep needs to be explored.

Managerial assessment of maternal behaviour

Maternal behaviour has also been assessed in various breeds by shepherds using the criteria of mothering discussed earlier and the ewe's behaviour when the lamb was touched. Six breeds of sheep have been assessed and the results summarized in Table 4.1. The Border Leicester × Romney were considered the most attentive mothers and the Merino the least attentive while the Cheviot had a high percentage of sheep which deserted their lambs but then returned later.

Breed and birth weight have both been shown to influence whether a ewe will stay with her lamb when disturbed. Another influence is the frequency with which sheep are handled by man. Arnold (1985) has shown that when ewes and day-old lambs were challenged by a man with his dog, 15% of sheep handled twice a year deserted their lambs compared with 5% of those which had been handled every month. In the latter case the sheep became more habituated and perhaps less fearful of the stimuli presented by man and dog.

The genetic diversity which is clearly present between breeds and between individuals allows the potential for selection for mothering ability. Selection for high total weight of lambs at weaning which favours twins at the time of weaning has been used successfully in breeding programmes as a means of increasing the percentage survival of lambs. This selection has integrated behaviours associated with being a good mother.

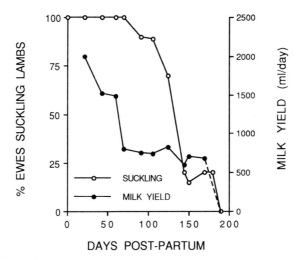

Fig. 4.20. The relationship between the percentage of ewes which allow their lambs to drink and milk available. Weaning occurred at about 800 ml/day for the Dorset Horn ewes (data from Arnold *et al.*, 1979).

Table 4.1. Lambing and behaviour results from breed groups in 1972 and 1973 (expressed as % of ewes which lambed in each group).

	Romney	Border Romney	Dorset Romney	Peren-dale	Cheviot	Merino
Dead lambs	9	3	5	5	6	15
Assisted	4	2	3	2	1	0
Mothering problems	7	3	4	3	2	1
Stood	52	78	82	59	41	11
Intermediate distance	37	19	16	33	28	38
Desert/return	5	2	2	7	14	28
Desert	5	1	1	2	12	23
Mothering						
Good	57	89	88	80	82	39
Average	36	10	11	19	13	36
Poor	7	1	1	1	5	26

Source: Adapted from Whateley *et al.* (1974).

Conclusions

A more complete understanding of the maternal behaviours observed in the parturient ewe has been achieved by studies of the relationships between the peripheral nerves and the central nervous system, together with the role some of the sex hormones and prior experience of parturition. There is more knowledge of the factors affecting the association between the ewe and its lamb. Several practical methods for fostering lambs have been worked out and better survival of the lambs has been achieved through strategic feeding and protection of ewes from adverse climatic conditions.

Further Reading

Alexander, G., Stevens, D. and Bradley, L.R. (1989) Fostering in sheep: an exploratory comparison of several approaches. *Australian Journal of Experimental Agriculture* 29, 509–12.

Arnold, G.W. and Morgan, P.D. (1975) Behaviour of the ewe and lamb at lambing and its relationship to lamb mortality. *Applied Animal Ethology* 2, 25–46.

Keverne, E.B. and Kendrick, K.M. (1990) Neurochemical changes accompanying parturition and their significance for maternal behaviour. In: Krasnegor, N.S. and Bridges, R.S. (eds), *Mammalian Parenting*. Oxford University Press, Oxford, pp. 281–304.

Poindron, P. and Levy, F. (1990) Physiological, sensory and experiential determinants of maternal behaviour in sheep. In: Krasnegor, N.S. and Bridges, R.S. (eds), *Mammalian Parenting*. Oxford University Press, Oxford, pp. 133–56.

5

Behaviour of the Lamb

This chapter discusses fetal behaviours which prepare the lamb for life after birth, and some of the responses to the huge array of sensory stimulation which suddenly assail the lamb at birth. The newborn lamb has no obvious means of recognizing its mother and yet it rapidly learns to seek her out, to find a teat and to suck. It must possess a series of innate behaviours and physiological adaptations to survive the dramatic transition from the steady-state environment within the uterus to a variable environment and a situation where it is essential to obtain colostrum and to follow its mother closely as she rejoins the flock.

Fetal Motility

There is a long history of interest in fetal activity of many vertebrate species, which has been generated in part by the clinician's need to help to ensure the well-being of the human fetus. In sheep, which have a gestation period of about 150 days, the classic studies of Barcroft and Barron (1937, 1939) have shown that the first movement of the fetus occurs in the fore-limb on day 34. Touching the nose of a 40-day old fetus results in a gener-alized movement of the neck, limbs and tail which is rapid and described as 'twitch-like' in character. By day 50 the movements are smoother and more prolonged. A few days later, stimulation of the nose results in the forelegs being extended and maintained in this position for some time together with an attempt to raise the head. This behaviour, like many subsequent reflex responses to stimulation of the five senses, is most important in the behaviourial sequences which result in the newly born lamb standing and seeking the udder. Between days 50 to 60, the fetus exhibits its first sustained muscular activity. Barcroft and Barron state 'it is difficult to escape the impression that the whole of the muscular system of the neck, trunk and limbs is concerned with an attempt on the part of the animal to orient itself with regard to space'. At this time, the position of the head appears to determine whether flexion or extension of fore or hind limbs as well as a generalized righting reflex occurs. The limb activity and righting reflex are initiated by head movement which stimulates various receptors within the neck.

A series of X-rays on the fetus from day 90 to 120 shows extension and flexion of the fore and hind limbs, the whole spine, and complex fetal movements ranging from the co-ordinated flexion and extension of limbs and skeleton to complete stretching of the whole body (Fraser, 1989). A full range of righting movements occurs which, in the pre-partum period, results in the lamb being in the birth position with its head near the birth canal. At this stage of development, the fetus performs rhythmic movements with its lower jaw which moves in both a vertical and horizontal plane, a probable precursor both to finding and sucking the teat and to selection and prehension of herbage.

By day 130, simple limb and neck flexion and extension occur throughout 24 hours with a frequency of 2 to 2.5 per minute or about 3000 a day. Fetal movements involving limbs and trunk are repeated several times during a bout of movement and occur 300 to 600 times a day. A further group of complex fetal movements involving the whole body are repeated for periods of 5 to 10 minutes. These episodes are seen as many as 15 times over 24 hours. Thus the fetus performs some 4000 to 6000 movements a day during the last two weeks of pregnancy.

The stimuli generating the intense fetal muscular activity related to righting, standing, sucking and swallowing remain unknown. It seems inescapable that the activity is associated with muscle maturation and tone so that the very vigorous activities of the precocious newborn can be achieved. These movements 'can equip an animal with a series of responses which appear ready-made at their first performance' (Manning, 1979).

Sensory System of the Fetus

The need for co-ordinated muscular and skeletal systems at birth is obvious. There is an equal necessity for a functional sensory system which has to receive a sudden barrage of novel external stimuli at birth. The sense of touch is well developed and a light touch on the face initiates the righting reflex which is needed immediately after birth. Within half an hour of birth this light touch is also used to initiate a completely different series of innate behaviours associated with the sucking reflex (Vince and Billing, 1986). Taste is well developed in the fetal lamb by day 100 and there is circum-

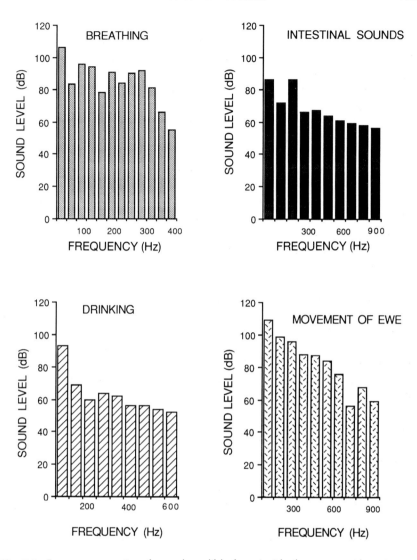

Fig. 5.1. Frequency spectra of sounds audible from inside the uterus with various activities of the ewe (data from Vince *et al.,* 1982).

stantial evidence for olfactory development associated with the recognition of amniotic fluid. Vision is almost certainly functional, but visual experience is unlikely before birth (Bradley and Mistretta, 1975). The fetal lamb has a functional auditory system which, for instance, receives sounds of the mother's feeding and movements of the alimentary canal. Some of the sounds audible within the mother have been recorded in Fig. 5.1.

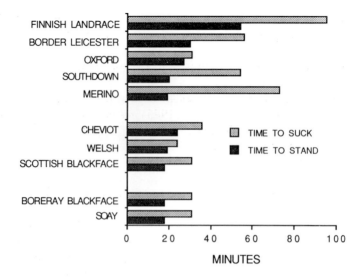

Fig. 5.2. Time taken for lambs of different breeds to stand and suck (data from Slee and Springbett, 1986).

The Lamb's First Hour of Life

The newborn lamb shakes its head and gives several inspiratory gasps prior to settling into regular respiration. Within a few minutes of birth the lamb shows minor movements of the head, neck and extremities. The righting reflex and subsequent standing, which was shown to be functional *in utero* is activated by the ewe licking the lamb's face and by the newborn lamb raising its head which in turn stimulates proprioceptive reflexes in the neck. The frequency and co-ordination of these movements increases dramatically over time so that usually within 60 minutes, the neonate stands and a well-organized and generally successful attempt at teat seeking and sucking occurs.

The time taken to stand and to reach the udder varies greatly both between breeds (Fig. 5.2) and within breeds (Slee and Springbett, 1986). Even though the Southdown and Merino stand quickly, they are slow in locating the udder. When speed and success in locating the udder were considered, there was a fairly clear distinction between the British hill and feral sheep on the one hand and the lowland breeds on the other. However, another study of seven lowland-type breeds showed no difference between breeds in the time lambs took to stand, approach their mother and finally suck a teat.

Providing there is no dystocia it is generally true that within a breed the heavier the lamb the more rapidly it stands and has its first suck. Lambs from primiparous ewes are slower to reach the teat than those from multi-parous ewes and multiple-born lambs are slower than singles.

Sensory system stimulation of the neonate

The presence of the mother provides a large range of stimuli to the newborn lamb. Research by Vince *et al.* (1985) has produced evidence that these stimuli are associated with the lamb standing, approaching its mother and teat-seeking. The obvious stimuli the mother provides relate to sight, sound, touch and smell.

Effects of stimuli for movement, sound, touch and vision

The effects of stimulating the lamb's senses of vision, hearing and touch on the time taken to stand for one minute have been studied by taking the lambs as soon as they were born and giving them one of seven stimuli set out in the following table (Table 5.1). Where the treatment required it, the operator moved the model ewe or food container or used a small paint brush to systematically touch the lamb over its whole body. In this study movement of head, fore and hind legs, standing, falling and several other behaviours were recorded on a checklist every 10 seconds for one hour so that the sequences and duration of those behaviours were quantified.

When first placed in the arena, lambs tended to make limb movements or righting movements, but later there were extended periods without activity. For purposes of analysis, quiescent periods were arbitrarily defined as more than one minute during which lambs made no limb movements at all.

Different treatments affected the proportion standing after 15 minutes (Fig. 5.3). The stationary model was associated more than any other group with quiescence. The stationary model group can be contrasted with the

Table 5.1. The seven stimuli offered to neonates which were kept isolated for one hour in a dimly lit room. The senses stimulated are also shown.

Stimulus	Sense stimulated
Moving model ewe	Vision
Ewe bleats	Hearing
Moving food container	Vision
No stimulation	Nil
Systematic light touch over body	Touch
Stationary model ewe	Vision
Ewe	Vision, hearing, touch, smell

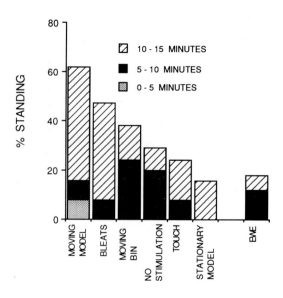

Fig. 5.3. The percentage of each treatment group which stood within 0 to 5, 5 to 10 and 10 to 15 minutes.

'no stimulation' group as it differed only in the presence or absence of the model. The amount of bleating fell in the presence of the stationary model during the second half of the test period whereas it rose in the 'no stimulation' group. Movements of righting and standing were lower in the 'stationary model' group, which also produced more quiescence. Thus the presence of the ewe model had a quietening effect on the lambs. Touch resulted in the lamb being relatively inactive unless it was directed to the perineal area. With the latter the lamb made vigorous attempts to stand. Low rumbling bleats were subsequently shown to cause quiescence in the lamb. As a result of the treatments the lambs fell into two main groups. 'Bleats' and 'no moving' groups were faster than the rest in the time taken to stand while the 'touch', 'ewe' and stationary model groups were slower.

On standing the lambs nosed the ground or, if in the model or the bin groups, they nosed these objects. The nosing consisted of a rapid back and forth movement of the head so that the muzzle touched the ground or the model. This nosing or bunting occurred soon after birth with the moving stimuli and stationary model (Fig. 5.4). The treatments of touch, bleats and no stimulation resulted in little nosing activity. However, it is clear that movement *per se* had a marked stimulatory effect on the lamb which resulted in it approaching and nosing the model far more rapidly and frequently than if the model was stationary (Fig. 5.5).

Lambs that had not sucked tended to sit quietly in the presence of a large stationary object when they were not getting any other stimulation.

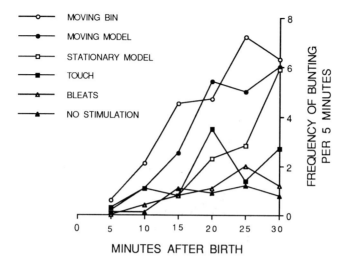

Fig. 5.4. Mean number of 10-second intervals in each of six successive five-minute periods of testing in which lambs nosed the ground or the model.

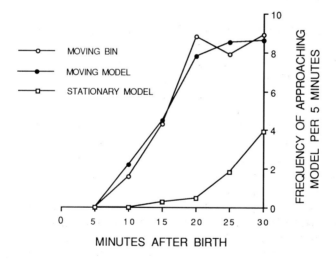

Fig. 5.5. Mean number of 10-second intervals in each of the six successive five-minute periods in which lambs from the three active groups approached the model and touched it.

This may well be similar to the behaviour of suckled lambs in the field. When the ewes move away from the birth site to graze or drink, their lambs often remain sitting or sleep quietly against alternative large objects such as a pile of straw bales or even in clumps of tall pasture. They even move towards such objects. This could well result in a reduction of mismothering as the mother may well be able to find her lamb more easily when she returns. This response to a large stationary object is one which requires more experimental investigation.

Stimuli from the mother

Clearly, the ewe must be regarded as a composite stimulus for the newly born lamb. She presents as a large visual object, which can be stationary or mobile, provides tactile stimulation and makes low rumbling bleats. When these aspects were presented separately they had different effects on the lambs. The results suggest that the ewe, as a large visual object, initially slows the lamb's progress to standing by being stationary and by tactile stimulation and bleats. Although this occurs, the ewe's presence and behaviour, particularly the licking of the lamb's hindquarters, appear to keep the lamb moderately active. By slowing the lamb's movements initially, the ewe lengthens the time during which the lamb becomes acquainted with her odour and any other characteristics. At the same time, by reducing its early struggles to stand, the ewe may help to conserve some of the lamb's initial meagre energy resources.

The dam licks the lamb's perineal area much more than other areas during the first second and third quarter hours of its life. This is a time when the perineal licking stimulates the lamb to enhanced activity associated with standing, bunting and seeking the udder. By this time, the ewe is more active which also stimulates behaviours associated with teat seeking.

Stimulation to suck

The innate behavioural responses of the lamb when its senses are stimulated during the time between standing and seeking the teat, have been studied in a classical series of experiments. This work has been reviewed by Vince and Billing (1986) and summarizes a series of papers written by Vince and her colleagues in the mid-1980s. The research is an excellent example of an ethological and experimental approach to a problem which is not often used in studies on domesticated animals. The lambs used in these experiments were removed from their mothers as they were being born. Lambs respond to temperature, odour and touch by a series of behaviours which must be innate initially and that change rapidly once the lamb has sucked.

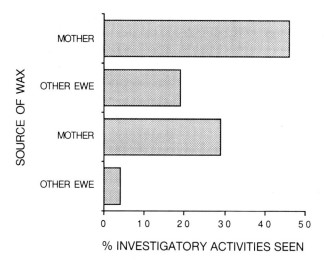

Fig. 5.6. Mean amount of lamb activity associated with odour sources from wax, which were derived from mother or alien ewe. The activities were investigatory in type, for example, following the stimulus, head movement towards the stimulus, mouth opening and bleats (data from Vince and Billing, 1986).

Temperature Lambs move up an increasing temperature gradient such as that measured on the ventral surface of a ewe. This 20°C gradient from the legs to the anterior/posterior ventral surface of the ewe means a lamb was inevitably attracted to the area between the shoulder and the brisket or the inguinal and udder depending on which way it moved when it touched the ewe. The thermal responsiveness of the lamb leads it to the area that results in its finding the udder and the teat through other innate behaviourial mechanisms.

Odour and touch Responses to odour and touch were observed as a vigorous forward and upward movement of the head, movement of the lips and tongue, and protrusion of the lips resulting in the lips and tongue forming a circular opening. By observing these responsive behaviours it was possible to compare a lamb's activity in response to its own mother and to alien wax. The lambs were stimulated to investigate and preferred mother's inguinal wax. The results are illustrated in Fig. 5.6. Lambs were also found to prefer smooth to woolly surfaces and were yielding to hard surfaces. Fig. 5.7 illustrates the sensitive areas of the unsuckled lambs. Movement of the head upward (orientation or forward) and opening of the mouth could be elicited more or less frequently depending on which part of the head was being stimulated (Fig. 5.8). Touch on the forehead, eyes and nose elicited more orientation, while touch on the mouth elicited more oral activity than other sites. These reflexes are lost once the lambs have sucked.

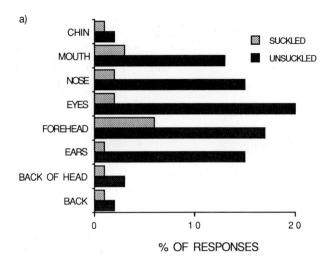

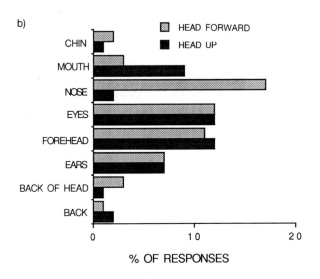

Fig. 5.7. Amount of oral activity elicited from unsuckled and suckled lambs by tactile stimulation at different sites (a), and orientation (head up, head forward) (b). Touch on forehead, eyes and nose elicited more orientation, while touch on mouth and nose elicited more oral activity than other sites (data from Vince and Billing, 1986).

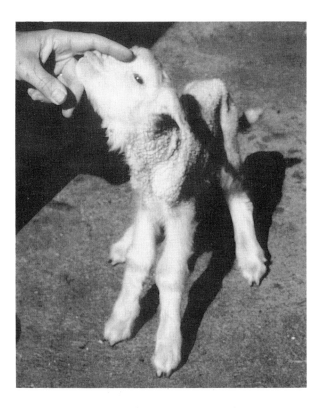

Fig. 5.8. Touch on the forehead and nose results in the unsuckled lamb lifting the muzzle upwards with the head coming forward.

Successful Udder Seeking

The responses of the unsuckled lamb, then, induce a series of behaviours which result in the lamb finding the teat and sucking. The lamb moves up a temperature gradient from the belly to the front or back legs. When its movement is towards the rear it is further attracted to the inguinal area by its mother's wax and to the udder by the smooth yielding surface.

Licking the lamb's perineal region stimulates the lamb's neck and head to move forward as well as increasing bunting and movement of the lips and tongue. As the lamb touches the belly with its forehead and eyes the head goes up and forward. When the nose/mouth touches the udder and teat, intense oral activity is stimulated. When the lower jaw/lip is touched there is movement of the head and mouth opening in the direction of the stimulus. When the teat touches the jaw and lip it will inevitably be seized. When the tongue is touched it causes closure of the mouth, often with the

Fig. 5.9. Sucking behaviour of the young lamb.

tongue curled around the stimulus, together with chewing and swallowing activity. Sucking is accompanied by rapid wagging of the lamb's tail (Fig. 5.9).

The importance of these behaviours being triggered in the newborn was shown by Vince *et al.* (1987). When the upper lip of a lamb was anaesthetized by infraorbital nerve block, the lamb still found the udder, but the stimulus for oral activity was lost and the mouth slid down the udder with the head appearing between the back legs. As the effects of the nerve block wore off the lamb sucked.

Many of the reflex behaviours are lost after the lamb has sucked. In fact, a lamb which has sucked can easily be identified since it loses all the behaviourial responses previously associated with touching various areas of its head and face. When a lamb which has successfully sucked from the teat is touched over the eyes and nose area that lamb no longer possesses the head-up and mouth-open reflex seen in the unsuckled lamb.

The teat-seeking reflex is not entirely driven by hunger since it is not completely suppressed by giving the lamb large quantities of colostrum. Teat-seeking, however, is markedly depressed by a drop in the lamb's rectal temperature to 37°C or lower as a result of cold wet weather (Fig. 5.10); a situation likely to occur when the neonate experiences wind and rain in the field.

The well co-ordinated series of interactions between the multiparous ewe and its lamb which culminates in sucking the teat can be short-circuited. Lambs which were blindfolded at birth remained motionless for the period

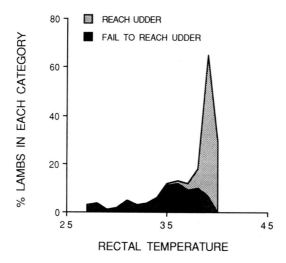

Fig. 5.10. Frequency (%) distribution of rectal temperature measured about one hour after birth for lambs failing or succeeding in reaching the udder (data from Slee and Springbett, 1976).

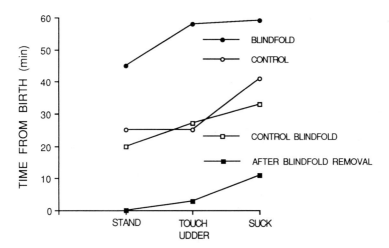

Fig. 5.11. Mean time for lambs to stand for one minute, first touch the udder and suck.

of blindfolding despite the presence of auditory, olfactory and tactile stimuli from the mother. After removal of the blindfold the lambs immediately stood and many moved rapidly along the flank to the udder and sucked rapidly (Fig. 5.11). Similarly, lambs which were removed at birth and placed blindfolded in a stimulus-free area for an hour before removing the blindfold and returning them to their mothers sucked within 15 minutes.

All these observations suggest that in some way the lamb is programmed by 'instinct' to move towards some kind of visual configuration. Further experimental work is clearly needed to clarify the characteristics of this visual configuration.

What is the purpose of the well-programmed teat-seeking behaviour when observations (Fig. 5.9) suggest these interactions between the ewe and her lambs are not absolutely necessary for successful teat-seeking and sucking? Suggestions have been made that this close interplay between the ewe and its lamb which involves several senses could be important to subsequent bonding or attachment.

Relation between sucking frequency and time after birth

Once the lamb has suckled for more than 60 seconds it has learnt how to find the teat and even if removed from the ewe for an hour and placed in a stimulus-free environment the lamb, when released, will find the teat within one to two minutes.

The frequency of sucking is high on the first day (Fig. 5.12) then declines to about once an hour by the end of the first week. Thereafter it progressively declines to about once every three hours by the ninth week of age (Hinch, 1989). There is very little difference in sucking behaviour between lambs of different breeds. After the ewe briefly investigates and identifies her lamb, sucking commences. Later, it is stopped by the ewe. With increasing age, and presumably more efficient sucking with experience, the duration of sucking is reduced.

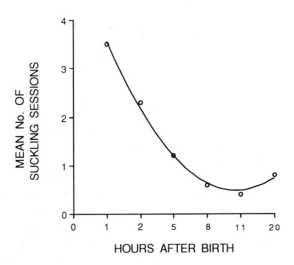

Fig. 5.12. Number of suckling sessions by the lamb in 15-minute periods at intervals in the first day of life (data from Bareham, 1976).

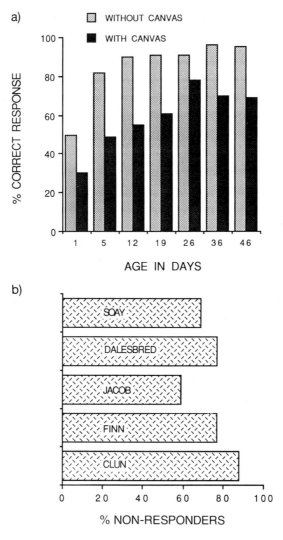

Fig. 5.13. The percentage correct response made by lambs in recognition trials related to age (a) and (b) the percentage of lambs of different breeds not responding to bleats coming from behind canvas (data from Shillito, 1975).

Increase in litter size results in an increase in sucking frequency at all ages compared with singles. From birth, multiples suck the teat at will but after the first week they are allowed to suck only if they are together. When the mean percentage of sucking bouts missed was measured for each twin and triplet it was found that a twin might miss up to 20% and a triplet 40% of sucking bouts which could contribute to higher rates of mortality in twins and triplets.

Recognition by the Lamb

Of the mother

The ability of lambs to recognize their mothers is clearly affected by the distance between them and by the age at which they are being tested. There was a marked improvement in ability to recognize the mother at a distance over the first four weeks of the lamb's life. When they were placed 10 m apart, most three-week-old lambs found their mothers rapidly and similarly most eight-week-old lambs recognized their mothers at a distance of 30 m (Hinch *et al.*, 1987). The speed of recognition of dams by their lambs appears to depend on breed and litter size. More twelve-hour-old Border Leicester × Merino than Merino lambs recognized their mothers when they were close together no matter whether they were born as singles or twins (Nowak and Lindsay, 1990). It took up to 24 hours before single Merino lambs could recognize their mothers and many twins took two to three days to do so (Nowak, 1988). Other research by Shillito and Alexander (1975) has shown that Clun Forest, Finnish Landrace, Jacob and Soay lambs could discriminate between mother and alien ewes within 24 hours of birth. When ewes were hidden by canvas, Clun Forest lambs were most and the Jacob least successful in recognizing their mother's voice (Fig. 5.13).

By the third day, Clun Forest, Jacob and Dalesbred lambs could all recognize and run to ewes of their own breed even though the mother was not among the ewes present during the test. At a distance, it appears that lambs from three days onwards can recognize their mothers by vocalizations, but they recognize the identity of the breed by sight. One characteristic of a lamb standing near an alien ewe of the same breed was that it continued to bleat. This did not occur when it stood near its mother.

Life-size models of Dalesbred and Jacob sheep in which the head moved and bleated with own mother's bleats were tested for recognition by three-week-old lambs. Both breeds approached the models on the first test, but only half the lambs moved towards the models the second time (Shillito-Walser *et al.*, 1985). The evidence suggested that the lambs responded to the bleats, but not to the movement. Clearly, the lamb viewed the model or its movement differently to the way the experimenter saw Dalesbred and Jacob sheep when the models were designed.

Of siblings

Association between siblings has been studied after transplantation of embryos so that ewes of Jacob and Dalesbred breeds were carrying one Jacob and one Dalesbred fetus. Over the experimental period there was a much higher level of association between siblings of different breeds than with aliens indicating that genetic inheritance has little to do with sibling

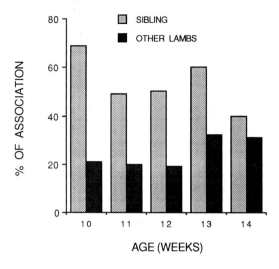

Fig. 5.14. Percentage of times that embryo-transplant lambs associated with sibling or other lambs (data from Shillito-Walser *et al.,* 1981).

recognition. Figure 5.14 illustrates this high rate of association between siblings. In young lambs the association appears to be made through the sense of smell and vision, while bleats in older lambs become a more important aid to recognition.

It is important that future studies focus on twins and triplets from birth to determine the relative importance of the dam and the sibling in these associations. In some breeds, the recognition of both the mother and the sibling when the mother moves away may well decrease the chances of a lamb being deserted during the first few days of life.

Importance of Interactions Between Mother and Young

In successful teat seeking

Behaviourial research on lamb survival has mainly involved the parturient and maternal ewe rather than the lamb although French and English workers (Poindron *et al.,* 1980; Vince and Billing, 1986; Nowak, 1990) have emphasized the behaviourial responses of lambs and the ewe/lamb interaction. Little attention has been given to the interactive behaviour of the mother and her young over the first few hours of the lamb's life.

The ewe induces a series of responses in the lamb as it stands and attempts to find the udder. The lamb's activity itself clearly stimulates maintenance of maternal behaviour. For the ewe's part, its behaviour

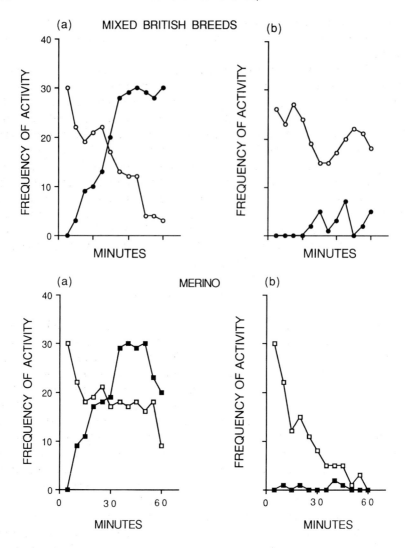

Fig. 5.15. Frequency of licking (ewe) and standing (lamb) during the first hour of life. (a) All lambs were active. (b) Lambs were blindfolded and inactive.

changes in response to the lamb moving in under the flank. Interactive behaviour is seen as the lamb approaches the udder. It pushes against the ewe's belly and udder, with the result that the ewe arches her back and pushes out a hind leg, exposing the teat. In doing this the ewe stops circling, stands still and takes up the position associated with suckling. Her head is turned back and she noses and licks the lamb's perineal area. Touch in this area is known to increase the lamb's nosing, bunting activity and tail wagging. The net result of these interactive activities is an intense drive of the lamb to seek the teat. Once having sucked the lamb often lies down quietly.

The interaction between the ewe and its newborn can be illustrated by an experiment in which there was temporary interference with some of the lamb's special senses (Vince *et al.*, 1987). Anaesthetizing the upper lip interfered only with the ability of the lamb to grasp the teat and suck, and depriving lambs of olfaction had no observable effect. Lambs which were blindfolded lay motionless for long periods regardless of breed. The behaviour of ewes which were Merino or a mixture of Dorset Horn, Corriedale and Suffolk is illustrated in Fig. 5.15. There were no differences in the ewes' behaviour provided the lambs were active. When the lambs lay still, licking by Merino ewes declined to zero over the second half hour of the lamb's life and interspersed grazing with licking until grazing was the predominant activity as maternal behaviour waned. Maternal behaviour waned more slowly in the 'British' breed.

In successful bonding or attachment

In the literature on the relationship between the ewe and her lamb the term bond is frequently used. In his review on mother/young attachment, Gubernick (1981) suggested that there is a distinction between a bond, which implies a more or less mutual, consistent and enduring tie, and attachment behaviour which may change as a function of the situation. The behaviour between ewes and their lambs suggests that attachment and not bonding is the appropriate term as the behaviour of the ewe and the lamb changes. This is consistent with the concept of maternal attachment being largely functional in maintaining care of one's *own* young rather than another's young and also with lamb attachment serving to ensure resources being provided by the mother. Initially these resources are nutrients, but soon after they involve learning of foods and habitats.

The attachment between the ewe and its lamb is strong after the first few days of the lamb's life and is maintained by the continued maternal care given by the ewe and the following behaviour of the lamb. In most, if not all domestic breeds, lambs follow their mothers or any other moving object. This is in contrast to many other ungulate genera where the young remain hidden (hiders) and do not move while the mother grazes. However, since goats have been observed to be either followers or hiders depending on the physical environment it is possible that sheep, a closely related species, may also adapt to be hiders or followers. Recent studies have shown that day-old lambs may travel up to 2 km following their mother and that the distance travelled is positively correlated with the strength of the association established on the first day of life.

Thus the mother/lamb behaviour is perhaps best seen as part of a continuum of different attachment behaviours observed in sheep flocks. Other influences may include sibling pairs, breed preferences and attachment to peers. These attachments are described in Chapter 2 on social behaviour.

Mechanism for ewe and lamb to find each other

A mechanism must exist for separated ewes and lambs to find each other. At a very early age the ewe recognizes its lamb's bleat and will seek it out. Lambs up to three days of age are identified at close quarters by olfaction, particularly by the ewe smelling the perineal region and then allowing the lamb to suckle. The lamb recognizes its mother's bleat and is attracted to a large object. The more a lamb bleats at birth the greater the time it spends with its dam when offered a choice of dam or alien ewe at 12 hours of age (Nowak, 1990).

Over the first month of a lamb's life the ewe seeks out and bleats frequently for its lamb (Hinch *et al.*, 1987). Thereafter, increasing numbers of ewes approach their lamb and stop, turn and bleat loudly before slowly walking to the flock. After the first month the lamb appears to play an increasingly active role in maintaining the mother/lamb attachment. As the lamb gets older the ewe no longer looks for her young, but remains in the flock and calls.

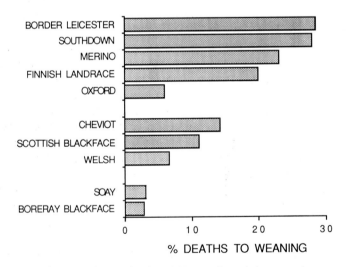

Fig. 5.16. Mortality rates among lambs of 10 pure breeds born on the same research station (data from Slee *et al.*, 1980).

Lamb Mortality

Most lamb deaths occur within the first three days of life and the reasons are many. This section reviews the causes of lamb deaths and offers management options which will reduce them. Management and care of sheep did not change much from the days of early domestication until the last 500 years when non-shepherding systems started to develop. The combination of minimal care of animals, year-round grazing, more intensive animal production from pasture and, more recently, the importation of sheep to countries with different environments has resulted in many lamb deaths.

Death rates of newborn lambs are high (15 to 20%) throughout the world. There is a considerable variation in mortality between breeds lambing in the same environment (Fig. 5.16). This mortality can be reduced to around 10% by placing parturient ewes in group housing and it is further reduced to around 2 to 3% by individually penning ewes. Alexander (1984) has reviewed the literature and has summarized a recurring theme of many scientists that 'starvation, cold exposure, difficult parturition and relatively low birth weight are four, often interrelated factors that are involved singly or in combination with most lamb deaths, and which are themselves affected by a variety of factors, especially by the level of nutrition during pregnancy'.

Vigour

In the literature on pregnancy, parturition, lactation and lamb survival there is a recurring theme of vigorous vs. 'weakly' or non-vigorous lambs at birth. Woolliams *et al.* (1983) state that 'if the "weakly lamb" can be regarded as a cause of death it is the single most important cause'. They also state that 'the descriptive term "weakly lamb" is an admission of ignorance since there is no apparent causative agent or physical effect. However, the existence of genetic variation in its incidence points to the possibility that physiological or metabolic factors may be found to underlie the condition'.

It is not only lambs of low birth weight which have lower vigour. There are large breed differences in immediate post-parturient activity. For instance, heavier lambs of the Oxford and the Border Leicester breeds were likely to be immobile for periods of up to 30 minutes after birth compared with feral or hill breeds.

The reasons for poor vigour in lambs are related to factors affecting fetal growth and problems associated with the birth process. So long as dystocia is not a problem within a breed, the larger the lamb the more vigorous it is. Vigour has been measured subjectively by the neonate's activity and has been ranked from good to bad, or associated with a measure of the time taken to achieve some behaviourial end, for example, to stand or to suck. However, Alexander *et al.* (1990) have shown that the speed with which

Fig. 5.17. An example of the following behaviour of triplets and quadruplets.

newborn lambs of Border Leicester or Merino origin stand and suck bears no relation to the lamb's ability to stay with its mother.

The lamb born with poor vigour, for whatever reason, will not only be more susceptible to climatic stressors but will also be at a disadvantage behaviourally since the close association between the dam and its lamb has been shown to be related to lamb activity which together with its bleating will stimulate maternal behaviour. A lamb will need to be vigorous to follow a ewe which does not show all aspects of maternal behaviour and has a strong drive to graze or drink. Perhaps a better measure of vigour may be the ability of a lamb aged one day or less to stay with its dam (Fig. 5.17).

There appears to be little known about the volume of colostrum a lamb sucks from its mother in the first 24 hours in relation to the volume within the mammary gland. The less-vigorous lamb, especially within a twin pair, may not obtain enough energy from the colostrum to ensure its survival over that vital 24 hours. In cool, wet weather for instance, there is insufficient energy within the lamb itself to survive these conditions for long. Fetal growth is dominantly under genetic control during the first half of pregnancy and there is minimum variability in growth. In the last half of pregnancy, nutritional substrates from the mother interact with fetal hormones, organ development and placental efficiency which result in great variability in fetal growth. However, our knowledge of the factors

affecting fetal growth and development are only superficial and frag-
mentary. Hence the real reasons for small lambs and lambs with poor
vigour being born must await more detailed research.

Practical solutions aimed at reducing lamb deaths

Housing

Perinatal deaths can be reduced by providing single pens or cubicles within
an animal house. This cuts out climatic and behaviourial causes of deaths
as the lamb is under little thermal stress and, given time, almost any ewe
will show sufficient maternal behaviour to allow a close association with its
lamb which can then suckle. However, this is an expensive approach to
reducing mortalities.

Shelter

In Australia, use of shelter systems which provide reduction in wind speed
over the whole lambing paddock have improved survival of single lambs by
8% and twins by 15% (Alexander *et al.*, 1980). The physical topography
of the paddock which stimulates isolation during parturition, and results in
ewes staying for long periods on the birth site needs further examination.
Graziers frequently know which paddocks have high and low lambing
mortalities but the reasons for these mortalities should be considered in
future studies of paddock and shelter design. Studies have shown that
newborn ungulates are attracted to large vertical surfaces and remain
lying nearby. This may be important when considering shelters within a
paddock. Since British breeds and Merinos have been shown to give birth
to lambs around fencelines, perhaps fencelines are more suitable sites for
development of shelters.

Genetic selection

Programmes which have been based on selection for the presence of ewe
and its lamb at weaning have had some effect on increasing the overall level
of lamb survival in succeeding generations. This approach suggests that the
lamb, assisted by its mother, has overcome the environmental hazards
experienced by sheep being born in the field. Another approach showed
that the heritability of cold resistance measured artificially by resistance to
cooling was 0.7 ± 0.25. Hence cold resistance could be increased by
genetic selection, but would promote birthcoat hairiness. However, there
may be other ways of promoting cold resistance related to follicle density
or changing the ratio of primary to secondary fibres.

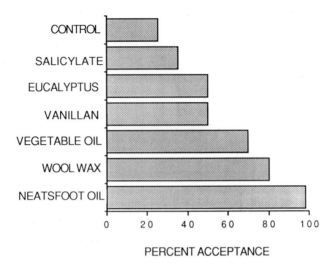

Fig. 5.18. Success rate of fostering odour-treated alien lambs (reproduced with permission from CSIRO, 1985).

Nutrition

Enhanced nutrition in the last few weeks before parturition increases birth weight and has been shown to decrease desertion of twin lambs. A 15% increase in survival of multiples compared with those of the control group was achieved by feeding ewes, during the second and third trimester, 80 g per day per fetus of a protein which is undergraded in the rumen (Lynch *et al.*, 1990). It is not clear why more lambs lived when ewes were given the supplement since it had no effect on ewe weight or lamb birth weight. Explanations may lie in the effect on placental growth, fetal metabolism or colostrum production. The supplement may have altered the immediate post-natal metabolism and thermoregulation, or changed the 'vigour' of the lamb so there was a stronger association with its mother. Increased vigour may have meant that the lamb was more successful in obtaining colostrum during the first 24 hours of life.

Fostering

Alexander *et al.* (1989) have examined methods of making alien lambs acceptable to mothers which could rear them. They found the age-old system described by Aristotle of skinning the dead lamb and placing the skin on the foster lamb plus confinement of ewes and the lambs was the most successful way of achieving fostering. When many methods of fostering were compared, it was found that confining the ewe and lamb and transferring odour from the ewe's lamb to the foster lamb with soft paraffin, detergent or a hessian coat was 80–90% successful after two days of confinement. There was variable success in the acceptance by ewes of lambs smeared with various commercial odorants (Fig. 5.18), with neatsfoot oil providing the highest acceptance rate.

Conclusions

The precocious nature of the lamb enables it to stand, suck and follow its mother for some distance four hours after being born. This chapter detailed the fetal preparedness for rapid post-natal development of the lamb as well as the innate responses of the lamb to touch, sound, vision and movement. The mother dampens some behaviours and enhances others so that the lamb stands after about 30 minutes. Although stimuli from the ewe affect the lamb's behaviour, even in their absence the lamb still stands and seeks the udder. Presumably this ability is needed for the lamb's survival in case the ewe deserts or dies. Various localized stimuli around the head and perineal region result in the lamb moving towards the ewe. Responses to ewe surface temperature, odour and skin texture result in the lamb finding the udder.

The interactions between the ewe and its lamb allow successful sucking and the formation of a strong attachment which is based on rapid recognition of each other. On the mother's part, this becomes exclusive to her own lamb. If the nutrition of the ewe has been good and there are no climatic extremes or rain, the union will survive until weaning. Methods of directing this knowledge towards improved management for lamb survival have been discussed.

Further Reading

Alexander, G. (1984) Constraints to lamb survival. In: Lindsay, D.R. and Pearce, D.T. (eds), *Reproduction in Sheep*. Australian Academy of Sciences and the Australian Wool Corporation, Canberra, pp. 199–209.

Nowak, R. (1990) Lamb bleats: important for the establishment of the mother–young bond? *Behaviour* 115, 14–29.

Vince, M.A. and Billing, A.E. (1986) Infancy in the sheep: the part played by sensory stimulation in bonding between the ewe and the lamb. In: Lipsitt, L.P. and Rovee-Collier, C. (eds), *Advances in Infancy Research*, Vol. IX. Ablex Norwood, New Jersey, pp. 1–37.

Vince, M.A., Lynch, J.J., Mottershead, B., Green, G. and Elwin, R. (1985) Sensory factors involved in immediate postnatal ewe/lamb bonding. *Behaviour* 94, 60–84.

6

Welfare of Sheep

Introduction

Interest in animal welfare is not a recent phenomenon. Concern for the well-being of animals has always been a central element in the relationship between people and domestic animals such as the sheep and still determines the success or otherwise of any system of animal husbandry. However, the focus of interest for animal welfare has changed in line with changes in farming practices. In the past, the major issue was direct inhumanity; that is, wilful or wanton ill-treatment of animals or that involving malicious intent. The scope of animal welfare now includes 'contingent' inhumanity (Hume, 1962); that is, inhumanity which might automatically attach to a system or procedure of husbandry. A catalyst for this change in scope was Ruth Harrison's *Animal Machines* (1964). A technical and socio-political landmark worldwide was the Brambell Committee which was set up in the UK in 1964 'to examine the conditions in which livestock are kept under systems of intensive husbandry and to advise whether standards ought to be set in the interests of their welfare, and if so what should they be'. Extensive systems of animal husbandry have also come under scrutiny; for instance, sheep husbandry in Australia (Senate Select Committee on Animal Welfare, 1990).

The change in scope of animal welfare and the use of modern husbandry systems has meant that the intuitive approach and the use of judgement and opinion based on 'commonsense' to identify and solve problems may no longer be adequate or sufficient. The demand now is for issues to be treated in clearly explainable biological terms and analysed through soundly based biological concepts. There are new problems to overcome and new challenges to meet. One problem is the belief that attention to the welfare of animals will impose economic costs and never benefits. However, there are sound physiological reasons for proposing a close connection between good welfare and good production performance of individual animals. The latter translates into good production performance by groups of animals. Accordingly, it is likely that attention to most animal welfare issues will confer benefits. Economic assessment of the costs and benefits of animal welfare is impossible without a mental picture of adequate welfare.

The theme in this chapter is the necessity for a scientifically defensible frame of reference for animal welfare which can be applied to particular problems in the husbandry of sheep. Nonetheless, intuition, opinion and judgement will still have a place in assessing welfare issues. Reasons for this are:

1. huge areas of uncertainties surrounding animal welfare;
2. intertwined philosophical and scientific problems which are still being identified and clarified before being addressed;
3. the need to act and make decisions about animal welfare at the present time.

However, the intention in this chapter is to continue the move towards a scientifically intelligible and defensible treatment of animal welfare problems and not to set out codes of practice or prescriptions on housing, feeding and transportation. For this reason, reference will be made to a conceptual model for the abstract idea of suffering. Suffering is considered the key feature of animal welfare, setting it apart from other issues such as the prerogative to kill and eat animals or to make use of their products. These other issues fall within the more general scope of bio-ethics.

Although much of the understanding of animal welfare depends upon the study of behaviour, a more complete picture comes from a consideration of physiology and pathology. These connections will be dealt with in this chapter. An underlying theme is the behavioural, physiological and genotypic fit between sheep and their environment and the way in which this fit may be perturbed by human intervention. Welfare concerns arise as efforts by animals to adapt to the range of environmental stimuli and factors are caused to increase.

The Background to Welfare Issues Relevant to Sheep

The human/animal relationship

The sheep is a domestic animal highly favoured by people. The species has a long and worldwide history of close cultural, religious and economic associations with a variety of human societies (Ryder, 1983). For this reason, sheep can be used to explore a variety of subjects which touch on animal welfare. One important subject is the perplexing biological nature of the human–animal bond. On the one hand, care-giving behaviour from humans is central to the human–animal relationship and is a key determinant of the prosperity of animal husbandry systems. On the other hand, the biological relationship between people and domestic animals such as the sheep is that either of predator and prey or parasite and host. Animal husbandry occurs only because nett benefit accrues to people. Benevolent emotions involved with care-giving have adaptive value for the success of humans as predators or parasites of domestic animals.

The apparent conflict between care-giving and exploitation in the human–sheep relationship should be resolved both ethically and psychologically. However, it cannot be considered in isolation from concerns for the general environment. Paradoxically, the solution may lie more in encouraging environmentally responsible pastoral forms of animal husbandry such as those involving sheep and less in becoming increasingly dependent on cropping and arable farming. Complete reliance on cropping raises additional ethical concerns and may not be an option for human welfare on a global level. Cropping alone is unlikely to be sustainable and will have consequences for the equity of future generations of people. It is likely to destroy habitat, obliterate populations of animals, and reduce the diversity of species.

One answer to the ethical paradox involved in animal-based agriculture may lie in exploring the ramifications of deep or transpersonal ecology (Naess, 1989). Transpersonal ecology 'finds value, not so much in the individual as in the system, be it an ecosystem or the biosphere as a whole, each with its "interests" in self-maintenance' (Birch, 1990). In this light, populations of sheep and grazing ecosystems involving both sheep and people are given intrinsic value and become subjects for human obligation and stewardship.

Animal welfare and the behavioural sciences

Animal welfare is a complex subject with philosophical, cultural, historical and scientific aspects. It is intimately associated with the subject of animal behaviour and for two reasons. Firstly, behaviour is the prime indicator of states of welfare and improved knowledge of animal behaviour will provide the basis for improved diagnosis of problems. There is an intimate two-way connection between behaviour and physiological state. Secondly, advances in understanding the nature of animal suffering must derive ultimately from the behavioural sciences. In this regard, the core subject matter of animal welfare emphasizes aspects of applied comparative psychology rather than ethology. The reason is one of definition. Suffering is an abstraction which involves the possibility of subjective feelings and states. Subjective states depend upon ideas of consciousness and awareness and this, in turn, invokes the question of mind as the 'seat of consciousness'. Finally, psychology is the science dealing with the mind and with mental and emotional processes. Ethology, on the other hand, is the study of behaviour with emphasis on proximate causation, that is the development and immediate control of behaviour, and ultimate causation, that is the adaptive value and evolutionary derivation of behaviour.

Philosophical deliberations have centred upon suffering, pain and distress as the definitive issues for animal welfare (Singer, 1976). These issues lie at the moving boundary between science and philosophy where science 'throws out challenges to philosophy but is frequently itself confronted with problems inescapably philosophical in nature' (Thorpe, 1962). Animal suffering involves subjective experience and this raises a conceptual problem. One view is that there may be no logically sound way for showing whether people, let alone animals, share similar feelings and experiences (Brain, 1962; McFarland, 1985). The idea of privacy is a central question for philosophical deliberations on subjectivity and subjective feeling and experience. Thus, the issue for animal welfare may be whether suffering is numbered among those private mental events which according to Ryle (1966) 'occupy a small and inessential place in the total range of mental phenomena'.

The behavioural biology of the sheep

The general biology of the sheep provides a broad context for understanding behaviour and those particular features which bear upon welfare practice. Sheep are ungulates and members of the subfamily Caprinae, the goat antelopes. They fit with the grazers as opposed to the resource defenders in this subfamily (Geist, 1984). Resource defenders such as the serow live in 'highly productive and diverse habitats that supply all their needs year round and can easily be defended against members of the same species'. By contrast, the grazers live in 'less productive and climatically more severe habitats', are 'highly gregarious and roam over large areas'. Two behavioural features of sheep likely to be pertinent to welfare are flocking behaviour and the observation that wild sheep live for most of the year in separate flocks of males and females.

Sheep and goats are apparently similar animals within the same tribe. However, sheep differ from goats in several important respects and the two must be considered entirely separate. They have blood type and chromosome differences. Chromosome numbers, however, may not be of ultimate importance because the genus *Ovis* contains species with diploid chromosome numbers ranging from 52 to 58. In general terms, 'goats are specialized for cliffs' whereas 'sheep inhabit the open rolling land close to cliffs. These differences arise from different ways of dealing with predators:

sheep escape by running and clumping into cohesive herds, while goats are more specialized in putting obstacles and terrain with insecure footing in the way of pursuing predators (Geist, 1984). Flocking behaviour varies within *Ovis aries* and appears to be much more pronounced in Merinos than in Scottish Blackface (Ryder, 1983). 'Flocking is so important to the animal that it is not surprising that separation from the group causes distress. Watts claimed that distress in the slaughterhouse is caused not by apprehension of death, but by separation from the group. He noted that the mass castration of ram lambs give little trouble, whereas clipping of the mothers causes consternation because removal of the fleece prevents the lamb from recognizing its mother by sight or smell' (Ryder, 1983).

Separation and the anxiety it causes is likely to be a predominant source of suffering by the sheep. Indeed, separation is a potent stimulus of the hypothalamic–hypophyseal–pituitary–adrenal axis in the sheep and leads to high plasma concentrations of hydrocortisone. For this reason, the results of physiological experiments made with isolated sheep are likely to be invalid in addition to contravening good welfare practice.

The connection between animal welfare and production performance

The relationship between adequate welfare and good production performance by animals has been canvassed since the mid-1960s. It is now a source of confusion because of current arguments about whether either individual animals or groups of animal should be the focus of attention. The answer is that groups are made up of individuals and that there are important social interactions. Accordingly, individual animals and the group as a whole should both come under scrutiny.

Events producing this confusion set the scene for re-evaluating the nature of the link between welfare and production performance. In the first place, the Brambell Committee (1965) rejected a positive and diagnostic link between production performance and welfare because of a desire to ensure that all possible sources and causes of animal suffering were dealt with. A similar idea was expressed by Dawkins (1980) who states: 'unfortunately productivity as it is measured need have no connection at all with the welfare of individual animals'. There are specific instances of this, for example pâté de foie and force-fed geese, where animal products are indeed pathological lesions and encroach upon biological fitness. However, these instances are exceptions rather than the rule. Ewbank (1988) describes abuse, neglect and deprivation as possible causes of suffering in animals and states that 'the stock keeper's traditional claim that his animals cannot be suffering because they are producing so well, is probably a good defence so long as he is only thinking of abuse and neglect'. Ewbank continues by saying that 'high production alone is not necessarily a full defence against the accusation of deprivation!'. In this regard, abuse refers to deliberate maltreatment of animals whereas neglect refers to maltreatment which results from idleness, ignorance or overwork. Both can lead to clear signs of ill-health and poor well-being. Deprivation occurs when animals are prevented from satisfying physiological and/or behavioural needs. A range of states may result from deprivation. Some may involve decreased production performance and suffering. Others may entail suffering and adequate performance and others still may result in no unconscionable suffering.

The connection between satisfactory welfare and good production performance appears to have been devalued because production alone does not measure all aspects of satisfactory welfare. On the other hand, it is clear that production performance is a diagnostic indicator of important and basic aspects of welfare such as adequate nutrition and the absence of disease. Included here are the diseases of adaptation which result from overwhelming environmental and psychosocial stressors. The situation can be put as follows: production performance for most animal husbandry systems is a necessary but insufficient indicator of satisfactory welfare. Production performance can be proposed as the major component of schemes aimed at quality control for adequate welfare.

Other components of these diagnostic schemes would be directed towards the remaining problems which may be associated with the deprivation of behavioural and/or physiological needs. In this way, the present approach does not deny the important consideration that 'well-being is an issue on the level of integration of the whole animal' and that 'this is the level studied by ethology and behaviourism' (Van Rooijen, 1990). Nor does it discount the possibility of comparing presumed suffering in productive animals with patterns of behaviour reminiscent of psychiatric patients whose suffering is known to be intense.

Definitions and Descriptions of Welfare

The treatment of welfare in this chapter hinges upon the concept of suffering as a subjective state. Earlier descriptions and definitions do not directly describe the absence of suffering as the essential element of welfare. Standard dictionary definitions put welfare as 'the state or condition of health, happiness or well-being'. Hurnik *et al.* (1985) describe well-being as a 'state or condition of physiological harmony between the organisms and its surroundings' with 'the most reliable indicators of well-being' as 'good health and manifestation of a normal (sic) behavioural repertoire'. The pioneering Brambell report (1965) described welfare as a 'wide term that embraces both the physical and mental well-being of the animal' and went on to say that 'any attempt to evaluate welfare must therefore take into account the scientific evidence available concerning the feelings of animals that can be derived from their structure and functions and also from their behaviour'. Broom (1986) describes the welfare of an individual as 'its state as regards its attempts to cope with its environment'. This last definition clearly states the importance for welfare of the fit between an animal and the environment but does not spell out the component of feeling or subjective experience. The problem of definition lies in dealing simultaneously with a mental state which must be satisfactory and a physical condition which must be satisfactory and stating that both are linked and both are prerequisites for good welfare.

Suffering and the components of welfare

The issue of animal welfare can be proposed to have two components. The first relates to a lesion or threat of a lesion to an animal. The term lesion commonly refers to the morphological changes associated with disease but has a broader meaning in the present situation. It covers metabolic, structural, functional and behavioural alterations and derangements due to injury. It extends to behavioural deprivation and includes the idea that unfulfilled behavioural needs, the ethological deficit, can involve unpleasant experience for animals. 'Injury' relates broadly to harm and damage rather than being restricted to physical rupture. However, it conveniently embraces those animal welfare issues that arise from surgical operations such as castration and tail-docking. The second component of welfare is the assemblage of consciousness, private experience, emotion and affect that allows for the mental phenomenon of suffering, the aversive affect which may come with or after a lesion. The term affect refers to 'the feeling element in mood or emotion'.

Suffering is summed up as a 'highly unpleasant emotional response usually associated with pain and distress' (American Veterinary Medical Association, 1987). Another description of suffering states that 'suffering occurs when unpleasant subjective feelings are acute or continue for a long time because the animal is unable to carry out the actions that would normally reduce risks to life and reproduction in that circumstance' (Dawkins, 1990). Both descriptions hinge upon feelings and emotions. '... feelings are completely private affective experiences. Emotion is their partially revealed behavioural expression' (Livingston, 1985). The term 'distress' appears in many treatments of animal welfare. A veterinary definition is 'a state in which the animal is unable to adapt to an altered environment or to altered internal stimuli'. Ewbank (1985) uses 'distress' in a similar way to describe the adverse consequences when an animal's coping response is exceeded by a stressor. Distress is given a much wider, perhaps all-embracing, meaning in the legislation related to animal welfare in the UK.

The Diagnosis of Welfare Problems and 'Stress'

Diagnostic schemes applicable to questions of animal welfare can be dealt with in the light of a suggestion (Ewbank, 1988) that the word 'welfare' may be replaced by the words 'health' and 'well-being'. An important condition was that 'health is more than the mere absence of disease and well-being is more than the absence of discomfort and distress'. The essential additional element was not spelled out but can be proposed as the complex of private mental experience, emotion and feeling. When feeling or affect is unpleasant and aversive it is identified as 'suffering'. The idea in this chapter is that analysis of welfare problems ultimately centres upon the presence and importance of the abstraction of 'suffering' in animals in given situations. Diagnosis depends upon an understanding firstly of how suffering may be incited and secondly of the preconditions in the animal which allow it to occur. In other words, a conceptual model for suffering forms the centrepiece in considerations of animal welfare. Species-specific knowledge for the sheep is essential if this model is to have application. The frequently quoted statement from Jeremy Bentham (1789) is highly pertinent and worth repeating in this setting. 'The question is not, can they reason? Nor, can they talk? But, can they suffer?'

Concepts of health and well-being contain common elements and are inextricably related. Unfortunately, these concepts are troubled by circular argument. For instance, health embraces the absence of disease. On the other hand, the presence of disease can entail discomfort and, therefore, poor well-being. Emphasis on disease allows a convenient evasion of the vexed question of 'stress' as an entity in its own right. This term has been used loosely and is likely to convey vitally different shades of meaning to different audiences and in different circumstances. The solution is to cut across much of the literature on stress and return to an older idea classifying 'stress' under the diseases of adaptation (Selye, 1946). Diseases of adaptation occur when mechanisms for coping with aspects of the psychosocial, nutritional or physical environment become overextended and break down. The general adaptation syndrome with its anatomical and functional consequences in the pituitary and adrenal glands and the immune system and the alarm reaction are only two aspects of diseases of adaptation. Other less understood interactions between body systems are likely to be important. Questions of welfare converge ultimately onto systems physiology and the network of regulatory pathways between body systems. For example, inter-regulation occurs between the immune system, nervous system and endocrine system. This particular set of relations has important implications for animal welfare and will be dealt with in its own right.

Another useful classification system for stress lies outside the present diagnostic scheme for welfare but helps materially in understanding concepts. Stress, overstress and distress are seen as part of a continuum of responses to a stressor (Ewbank, 1985). The term physiological stress applies when the response to a stressor occurs at a fully adaptive harmless level and the animal copes within its capacity. Overstress describes the situation where the coping mechanism is stretched but the animal is still able to handle the stressor. By contrast, distress occurs when the coping mechanism has been stretched beyond its limits and the response to a stressor is damaging and non-adaptive.

If health and well-being are taken as the primary components of animal welfare, and include the absence of diseases of adaptation ('stress'), they can be evaluated through the familiar diagnostic considerations of physical examination, production performance, behaviour and examination of the environment. If behaviour is made an important focus and its evaluation is systematized, the scope of diagnosis could be made comprehensive. Production performance can be emphasized for sheep. The important parameters of body growth, wool growth and reproductive activity all connect with general biological fitness in the ovine genotypes used by people in animal-based agriculture. However, some forms of husbandry of Merino sheep which involve varying degrees of undernutrition to deliberately produce extremely fine grades of wool approaching mohair may be an exception. This is unacceptable since the use of the appropriate genotypes can achieve the same end. As a particular example, treatment of selenium deficiency in Merino sheep at pasture will improve reproductive rate, body-growth and wool-growth but may decrease the overall profitability of a superfine wool enterprise by increasing fibre diameter. This phenomenon does not occur when fibre diameter is more completely determined by genotype. The solution to both the welfare and production problem is genetic selection after selenium treatment.

Disease and Welfare in General

There are important connections between disease and welfare (see Fraser and Broom, 1990). Basically, disease and welfare are linked in two ways. Firstly, the presence of disease can indicate a breakdown in adequate husbandry and point to underlying welfare problems. Secondly, disease itself may give rise to unpleasant affect or suffering and is a reason for direct welfare concern.

The first connection can best be explained by reference to definitions of disease. Blakiston's *New Gould Medical Dictionary*, for example, gives two definitions of disease both of which contain two elements. The definitions are as follows.

1. The failure of the adaptive mechanisms of an organism to counteract adequately the stimuli or stresses to which it is subject, resulting in a disturbance in function or structure of any part, organ, or system of the body.

2. A specific entity which is the sum total of the numerous expressions of one or more pathological processes. The cause of a disease entity is represented by the cause of the basic pathological process in combination with important secondary causative factors.

'Failure of adaptive mechanisms' in the first definition and 'important secondary causative factors' in the second can be regarded as related concepts. They form the first component in the idea of disease. 'Stimuli and stresses' in the first definition and 'the basic pathological process' in the second can also be regarded as related concepts. They form the second component in the idea of disease. It follows that the 'basic pathological process' or the 'stimuli and stresses' are a necessary but insufficient cause of disease. In other words, the appearance of disease depends on factors additional to the presence of the primary causative agent (a pathogen, metabolic dysfunction, deficiency, or chemical or physical stressor). For many diseases, especially those which have reduced production as their major or only manifestation, these additional factors, the secondary causative factors, are critical. This last point has far-reaching implications not only for disease control and therapy but also for welfare. Secondary factors can influence immune function and increase susceptibility to disease.

Figure 6.1 shows how primary and secondary causative factors interact with the adaptive or coping mechanisms in an animal's nervous, immune and endocrine systems to give rise to disease. The accompanying concerns for welfare relate to the disease itself. They also relate to deficiencies in husbandry which may underlie the secondary causative factor. This is the important second connection between disease and welfare. Given newer knowledge about the effect of stressors on disease resistance, the incidence of disease (especially disease which is primarily registered by lowered production performance) may be taken as a sensitive indicator of 'welfare' (Broom, 1989; Gibson, 1988). The reason for this should be treated in some detail for it is likely to be an important growth point in science applied to animal welfare.

Final expression of any immune response, including protective responses against production-limiting diseases such as parasitism, is determined by regulatory circuits within the immune system (Janeway *et al.*, 1985) and by interplay between the immune system, nervous system and the endocrine organs (Payan *et al.*, 1986; Bateman *et al.*, 1989; Jankovic, 1989; see also Oppenheim and Shevach, 1990). This is shown in Fig. 6.2 which reflects current interest in immunophysiology. Communication pathways between the nervous, immune and endocrine systems are inferred from observations that psychosocial and physical stressors have a crucial

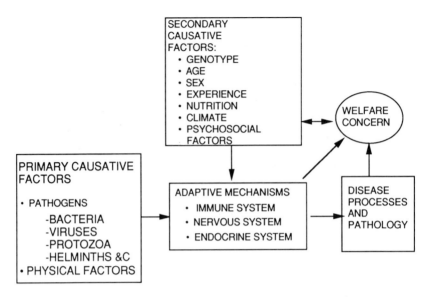

Fig. 6.1. Connections between the primary and secondary causative factors in disease, the adaptive mechanisms of an animal, the presence of disease and concerns for welfare.

influence on the course of infections and tumours (Riley, 1981; Plaut and Friedman, 1982; Kelley, 1980) and that behavioural conditioning can intensify or dampen experimental immune reactions (Ader *et al.*, 1987). The underlying anatomical and molecular connections indicate a reciprocal flow of information and control (Felten *et al.*, 1988; Blalock, 1988; Johnson and Torres, 1988). For instance, psychosocial stressors act through the nervous system to perturb both complex behaviour and immunity (Weinman and Rothman, 1967; Hamilton, 1974). Recent work with sheep has shown that an acquired immune response, that against the barber's pole worm, *Haemonchus contortus*, can act in the opposite direction and affect complex behaviour (Fell *et al.*, 1991).

As to the molecular mediators involved in the interplay between the immune nervous and endocrine systems, adrenal corticoids have long been known to interfere with immunity and are released during stressful situations (Selye, 1976). However, other molecules are influential. Stress-mediated immuno-insufficiency can occur after adrenalectomy (Keller *et al.*, 1983). Some of the neuropeptides may be involved. Accordingly, there is more to stress immunodeficiency than the action of adrenal corticoids such as hydrocortisone. It is also clear that substances produced by the immune system influence brain function and behaviour. Examples here are interleukin 1, a lymphokine produced in the immune system, which interacts directly with the hypothalamus to produce fever and indirectly

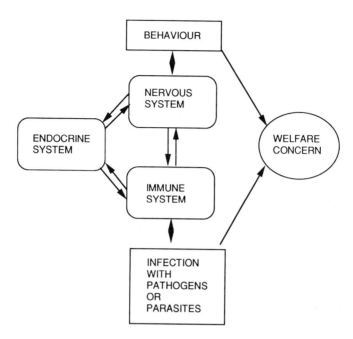

Fig. 6.2. Diagram showing the interplay between the nervous system, endocrine system and immune systems which results in the final expression of immunity and produces considerations for animal welfare.

decreases appetite and pain perception but intensifies slow-wave sleep (see Oppenheim and Shevach, 1990). The first point here is that molecules produced by the stimulated immune system could influence or even determine affect or feeling and may thus be the chemical mediators of suffering. The second point is that inferences about the connection between suffering and chemical mediators could be made through observations of behaviour.

Strong emphasis must be placed on the use of disease to determine the presence of environmental stressors which operate as secondary causative factors. These stressors represent a threat to animals even in the absence of primary causative agents of disease. They are objects of welfare concern in their own right. For this reason, diagnosis of disease with production efficiency and adequate welfare in mind should identify both primary and secondary causative factors. A good case in point for the sheep is parasitic gastroenteritis caused by the trichostrongylid parasites *Haemonchus contortus*, *Trichostrongylus* spp. and *Ostertagia* spp. and the important idea of primary and secondary parasitosis in this situation. Secondary parasitosis refers to the situation where immunodeficiencies from whatever cause allow for parasitism whereas primary parasitosis refers to the situation of overwhelming parasitic challenge in the face of otherwise adequate immunity. In instances where primary causative agents of disease are always likely to be present (e.g. nematode parasites or the orf virus) the diagnosis of disease and adequate welfare may eventually come down to the identification of immunodeficiencies and the factors and stressors which produce them.

A Conceptual Model for Suffering

The model for suffering put forward in this chapter is based on considerations of psychophysiology. It is not meant to stand alone in the analysis of animal welfare problems. Rather, it should complement other models such as that of Dawkins (1990) which follows the paradigm of economics and refers to the 'canonical costs' which apply when an animal is obliged to adjust to a given disturbance. There are two parts to the present psychophysiological model. The components of the first part are shown in Fig. 6.3. The first part consists of the input to the nervous system which may come from signals from the internal or external environment relayed either by neural transmission or by direct chemical mediation. This represents the input from the lesion or threat of lesion mentioned earlier. The input signal embraces those concepts of welfare associated with health, well-being, and deprivation or ethological deficit. Positive as well as negative inputs are included in the idea that environmental factors may act either as stressors or necessary stimuli. The input signal is processed within the nervous system to produce the unpleasant affect identified earlier as suffering.

The second part of the conceptual model for suffering refers to transformation of the neural and chemical inputs to the brain in such a way that the mental experience identified as suffering may occur. The great difficulty here relates to the idea of privacy as it applies to mental experience and completely satisfactory knowledge may be an unrealistic expectation. A set of criteria, which can be continually upgraded as new knowledge and information appears, may be all that will be possible for practical purposes. These criteria are the subject of the second part of the model. They operate from the set of preconditions or prerequisites in anatomy, physiology and behaviour which can be expected to underlie suffering.

The tendency when discussing animal welfare is to state that where doubt occurs, its benefit should go to the animal. If the present conceptual model for suffering is considered and exposed to continual analysis, the areas of doubt may be located with greater precision and may themselves become subjects for scientific as well as philosophical investigation. The notion of a moving boundary between science and philosophy and a renovated partnership between the two activities are both crucial for progress in animal welfare.

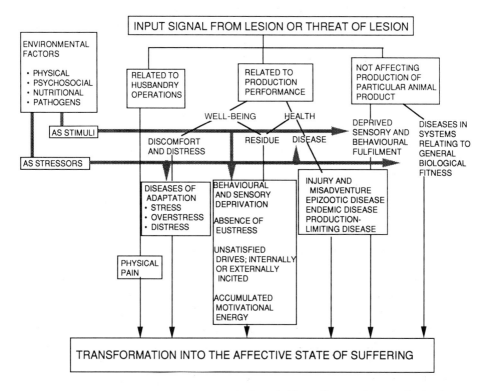

Fig. 6.3. The first of two parts in a conceptual model for suffering. This part deals with the input which is transformed in the nervous system to produce the unpleasant affect of suffering.

Major elements in the second part of the model for suffering are: (i) the psychological concept of feeling or affect; (ii) motivational state; and (iii) the necessity for animals to display the sort of adaptability in their behaviour that arises from the capacity to learn and is supported by the processes of memory and cognition. The model depends heavily upon the triad of thinking (cognition), feeling (affect), and doing (conation) that once formed the foundation of psychology. The triad is shown schematically in Fig. 6.4. An important consideration is that inferences about private mental experience, that is about cognition (thinking) and affect (feeling), can be made from observations of circumstance and behaviour. Behaviour in this setting can include emotion which has been described as 'the partially publically revealed expression' of 'completely private affective experiences'.

For purposes of the model, the term motivational state encompasses the concept of drives and their satiation, together with ideas about innate releasing mechanisms, releasers and accumulated motivational energy. This

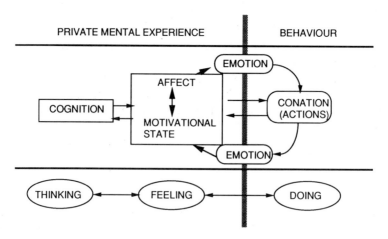

Fig. 6.4. The links with behaviour and private mental experience for the triad of thinking (cognition), feeling (affect) and doing (conation) which are ingredients of the second part of the conceptual model for suffering.

provides a platform for explaining why adaptability of behaviour and its connection with learning, memory and cognition is an important consti- tuent in the psychophysiological model for suffering. 'Drives unamplified by emotion do not activate learning or sustain behaviour' (Davidson, 1986). Emotion, affect and motivational state may form an indivisible whole when complex and adaptable behaviour are involved. Indeed, motivational states may arise when drive is compounded with emotion and affect. By contrast, there may be no necessity to involve emotion or affect in explanations of behaviour which occurs as fixed action patterns and which remains uninfluenced by experience or learning.

Physical pain related to tissue damage is a good example of how aversive affect or suffering is inherent in a motivational state and may actually govern behaviour. Wall (1979) has described the sequence of pain-related behaviour and has described pain as a 'need-state' similar to hunger or thirst. The behaviours associated with first or rapid pain, which arises from the threat of tissue damage, are linked to escape and avoidance whereas those associated with second or chronic pain, which arises from tissue damage itself, are linked to repair and recuperation. Unhappily, pain- related suffering frequently occurs when there is no chance either of escape and avoidance or recuperation and repair. It is clear from this that the precondition for suffering which involves the processes of learning, cogni- tion and memory, relates to the whole system rather than to individual circumstances.

Two other ideas add to the present psychophysiological model for suffering and the description of suffering as an unpleasant and aversive affect or feeling. First is the four-sided view of emotion and affect which is currently influential in ethology (Plutchick, 1986).

1. Emotions are considered 'within an evolutionary context as adaptive classes of behaviour'.

2. Emotions relate to 'complex chains of events that include more elements than expressive behaviour alone, or subjective feeling' but connect as well to 'cognitive appraisals, states of physiological arousal, impulses to action, and overt behaviour'.

3. The properties of emotions can only be known through the process of inference.

4. 'Emotions often mix and interact thereby producing various indirect derivatives'.

Second are schemes for the classification of affect and emotion. Affect for people has been divided into three categories which clearly connect with motivational state (MacLean, 1970). These could be applied to animals in given situations. There are basic affects, specific affects and general affects. Basic affects relate to hunger, thirst, sexual need, and the urge to defaecate, urinate and breathe. It is easy to imagine how thwarted behavioural and physiological needs may lead to suffering in animals under circumstances involving basic affects. Specific affects are related to the activation of specific sensory systems and include physical pain which, in terms of sensory physiology, is incited by noxious stimulation of specialized neural receptors. General affects include love, anger and disappointment and may be troubled by the worst in anthropomorphic thinking if they are extended uncritically to animals.

Another scheme for emotion in animals has been set out by Plutchick (1980) and could be the starting point for further work in the area. In this scheme, there are 'eight basic behaviour patterns which are the prototypes of emotions as seen in higher animals'. The terms for these behaviour patterns and their associated affect are as follows: (i) self-protection and fear; (ii) destruction and anger; (iii) incorporation and acceptance; (iv) reproduction and joy; (v) reintegration and sadness; (vi) orientation and surprise; (vii) rejection and disgust; and (viii) exploration and curiosity. Some of these behaviour patterns and their associated affect sound obscure. Some sorts of affect do not appear to be included but this is covered by the qualification that patterns 'may interact to produce the large variety of emotional states'. Perhaps frustration, 'the state of motivation that occurs when an animal's actions do not lead to the expected conse-quences or rewards' (McFarland, 1986) derives from interacting behaviour patterns.

Obviously the motivational states which may predominate for the sheep and provide the basis for affect, unpleasant or otherwise, will depend upon species-specific behaviour and its underlying evolutionary significance.

These in turn connect with the biology of *Ovis aries* and its environmental adaptations and will be described later in this chapter.

Pathophysiological aspects of welfare and suffering

Physical pain deserves particular attention in relation to the psychophysiological model for suffering. For convenience of explanation, physical pain can be separated into three components (see Adams, 1988). First is nociception or the detection of noxious stimuli by neural end-organs, the nociceptors. Second is the perception of noxious stimuli and the processing and integration of sensory signals within the central nervous system. This component has been designated as pain (Adams and Martin, 1983). The third component is suffering or the aversively unpleasant effective–emotional aspect of the phenomenon. The overall phenomenon of pain has two categories. These are first or rapid pain and second or chronic pain. They have different characteristics related to sensory inputs, to integration and processing of these inputs and to the adaptive value of the behaviour they produce. To clarify, first pain is produced by the threat of tissue damage and connects with escape and avoidance behaviour whereas second pain is incited by tissue damage itself and connects with repair and recuperative behaviour. These two categories have clear-cut implications for animal welfare. At present, first pain is given much more lay attention in animal welfare than second or chronic pain, but the latter is likely to cause more suffering in more animals and, by definition, is likely to last longer.

The question of pain has connections in the subject of pathology, firstly because its inciting cause is tissue damage or the threat of tissue damage and secondly because it goes with heat, redness, and swelling as a key component of the inflammatory response which occurs with injury. It is relevant here that pain is not a phenomenon completely dependent on electrical events in the nervous system but is additionally regulated by a family of chemical mediators which are involved in cell-to-cell communication. This line of thought is leading to the psychophysiological nature of other unpleasant feeling tones or affect which may qualify as suffering.

Chemical mediators are described as cytokines or lymphokines when they are produced by the immune system and neuromodulators or neuro-hormones when they are produced by the nervous system. All seem to be involved in changes of state in the central nervous system and in mediating the functional connections with the endocrine and immune systems. One set, the endorphins, are important regulators of pain, especially at the level of suffering. A lymphokine, interleukin 1, has direct effects on the hypothalamus to induce fever, a pathophysiological change with adaptive value in the control of disease. In people, fever is accompanied by unpleasant feeling tones which are also likely to be incited by chemical mediators.

This line of thought now terminates on the wide range of unpleasant feeling tones or affect which accompany disease of various sorts in people and which could also apply to animals such as the sheep. For people, these unpleasant affects include pain, malaise, nausea, fever, headache, depression, dizziness, fatigue and respiratory distress. Many of them are likely to have chemical mediators, either cytokines or neurohormones, as their inciting or proximate cause. Furthermore, the motivational states they are associated with are likely to have an adaptive value or distal cause according to the usual scheme for explaining behaviour. Malaise, for example, is associated with inactivity and is the feeling that brings about this form of behaviour. The result is conserved bodily resources and time for repair and recuperation. Repair and recuperation may thus represent the adaptive value of behaviour induced by malaise. It is almost axiomatic that disease of all sorts will be accompanied by changed motivational states and that these, in turn, are likely to involve unpleasant affect or suffering.

The wide range of unpleasant feeling tones and the different psycho-physiological sources of suffering raises many questions for a ruminant such as sheep. For instance, the question of nausea has been raised in Chapter 1 when conditioned feed aversions were dealt with. Nausea may occur in a ruminant such as the sheep which does not overtly vomit. Does vomiting occur in an obscure form such as reflux between the rumen and abomasum? Is this an accompaniment of sea-sickness/motion-sickness and does it occur during periods of transport at sea or on land? Questions such as these could guide the analysis of welfare problems linked to the important issue of transportation.

Behaviour and the Diagnosis of Welfare and Suffering in Sheep

Examination of behaviour is crucial to the diagnosis of problems in animal welfare. Two basic features about behaviour are relevant here. Firstly, there is a two-way connection and dependency between overt behaviour and the underlying physiological state. One familiar example of this two-way dependency is the connection between feeding–foraging behaviour and the physiological mechanisms that give rise to hunger. Secondly, behaviours have both a proximate and ultimate cause. Proximate causes of behaviour relate to the stimuli which trigger it and the part played by the nervous system, endocrine glands and musculature in expressing it. Ultimate causes refer to the long-range adaptive significance of the behaviour. Behaviour bears upon animal welfare in three ways.

1. Thwarted expression of behaviours may be a cause of suffering.
2. The study of behaviour may shed further light on the nature of suffering. Suffering associated with motivational states such as fear or frus-

tration fall within the experimental scope of comparative psychology.

3. An understanding of repertoires and patterns of behaviour of animals under various conditions and in a range of physiological states may provide a point of reference from which possible deviations may be assessed. This raises the question of abnormal behaviours in sheep and their welfare implications which will be treated separately.

The third point about behaviour, namely its application as a diagnostic instrument, should excite renewed interest in the preparation of ethograms for sheep of different breeds and in different situations. Ethograms are defined as the complete behavioural vocabulary of a species, the listing of units of behaviour whose occurrence in various contexts and sequences can be used in principle to provide a comprehensive description of behaviour (Harré and Lamb, 1986). The construction of ethograms for the sheep may present more difficulties than are at first apparent. Nevertheless, the attempt may result in the establishment and evaluation of observational methods which can be incorporated into diagnostic schemes for welfare.

One other family of behavioural tests is relevant to the assessment of adequate welfare. These are the motivational choice tests which may complement and extend ethograms. They can be used to evaluate aversion and thus give an index of the unpleasantness of different husbandry procedures and situations. However, motivational choice tests have more potential than this. They may be valuable for analysing cognition and motivational state, especially when these are associated with or influenced by various stressors and disease. A prime example is the arena test developed for use in sheep (Fell and Shutt, 1989). This test is based on flight distance and has been used to determine the aversion produced in sheep by castration, tail-docking and the Mules operation for controlling blowfly-strike and to unravel the interactions between behaviour and immunity.

Although thwarted behaviour can be a source of suffering, the causal connection may not apply in every instance. However, there is another perspective. The excitation of behaviours through the group of interventions classified as environmental enrichment may unlock or facilitate physiological activity and benefit production performance. This clearly repeats the theme about the intimate connection between productivity and adequate welfare. Improved productivity has been recorded in domestic pigs encouraged through their environment to express the same range of behaviours as their wild counterparts (Stolba and Wood-Gush, 1984).

Preliminary studies have been made to test for the same possibilities in sheep (Stolba *et al.*, 1990). The question posed was whether some behaviours may be stifled in Merino sheep husbanded at pasture in homogeneous groups according to sex and age and in flat open and treeless environments. The necessary comparison was whether husbanding sheep in family groups in an environment 'enriched' with trees and other features might elicit a wider range or greater intensity of behaviours. The experiment showed that a great expression of behaviours occurred in the enriched environment and within the heterogeneous social groups. It is not clear whether these results are peculiar for Merinos or which set of conditions were associated with improved production performance. Moreover, the results have implications for understanding what constitutes good welfare in sheep. Did the behaviour of sheep in the bland environment represent low-grade depression, that is 'behavioural atony' (Fraser and Broom, 1990), or did it indicate comfort and ease? Did the behaviour of sheep in structured social groups and in an enriched and stimulating environment indicate greater well-being or did it represent that response to stressors required to keep sheep within a subjective state of comfort? Greater species-specific knowledge about the behavioural biology of sheep and about the influence of domestication is required for answers to this dilemma in interpretation. Insight into states of arousal and the stimuli for innate and acquired anti-predator behaviour in sheep will also be relevant. Measurements of visceral alarming reactions such as increases in heart rates of sheep in response to dogs, to isolation and to new companions have already been helpful in this regard (Baldock *et al.*, 1988) and should be among the suite of observations required to establish the ovine ethogram.

Pain-related behaviour

Pain-related behaviour in sheep requires particular treatment. Pain can determine or modify voluntary behaviour and involuntary behaviour associated with the autonomic nervous system. Wall (1979) suggested that a general pattern of pain-related behaviour follows traumatic injury in mammals. This general pattern is likely to apply to sheep but the details will arise from the peculiarities of behaviour of this species. It is highly likely that the sequence of pain-related behaviour is induced by tissue-damage and governed by the nervous and endocrine systems. This sequence will be described before treating accounts of pain in sheep.

First in the sequence is an immediate phase where activities for avoidance and defence take priority. The influence of pain on function, and presumably unpleasant affect or feeling, may be absent at this stage or displaced by other forms of arousal. These absences are attested to by instances of racehorses winning races on broken legs and by accounts from soldiers about their immediate responses to severe battle wounds. The emergency, alarm or 'flight–fight' reaction which results from the co-ordinated response by the sympathetic nervous system and substances released from the adrenal gland can be expected here. Problems arising during this phase include circulatory shock and heart failure of various sorts.

Second in the sequence is an acute phase, 'a transition between coping with the cause of the injury and preparing for recovery'. This phase is characterized in people by a combination of 'tissue damage, pain and anxiety' (Wall, 1979) and is not well described in animals.

Third in the sequence of pain-related behaviour is a chronic phase characterized by quiet inactivity which is regarded as 'the optimal tactic to encourage cure and recovery of damaged tissue' (Wall, 1979). This phase may not have received the attention it deserves in animal welfare. It provides a useful approach to considerations of post-operative pain.

Little can be said about specific pain-related behaviour in the sheep because little is known. It has been considered generally in schemes for the diagnosis of pain in domestic animals (Morton and Griffiths, 1985; Sanford *et al.*, 1986). Pain-related behaviour has been investigated in particular circumstances such as those related to castration, tail-docking and tooth-grinding (Denholm and Vizard, 1986; Shutt *et al.*, 1988; Mellor and Murray, 1989). However, there is insufficient information upon which to build a general description of pain-related behaviour in the sheep.

A Comparison of Welfare Issues in Intensive and Extensive Systems of Sheep Husbandry

Intensive and extensive systems of husbandry present different sets of welfare problems. Those confronted by sheep in intensive husbandry systems such as feedlots, or indeed some research programmes, arise from confinement and restriction in pens. In contrast, the welfare problems confronted by sheep at pasture relate to lack of shelter at critical times, fluctuations in the food supply and some of the husbandry operations required for management. Overall, however, extensive systems have positive advantages for the welfare of sheep. Sheep at pasture are allowed greater scope for the expression of behaviours and to fulfil their 'telos'; that is, the innate being or destiny as an individual of a species (Fox, 1986). Sheep at pasture are able to move within their living space to optimize physiological and behavioural comfort. Space allows escape from the psychosocial stress associated with dominance behaviour.

Welfare issues associated with intensive systems of husbandry

Confinement underlies most of the specific welfare problems of sheep kept in pens. Confinement restricts the expression of behaviours which make for the comfort of sheep in the short, medium and long term. Various space allowances have been prescribed as meeting the welfare requirement of sheep in pens. These allowances may be soundly based on the best of opinion and judgement. However, they beg the question of good welfare and shift the onus of responsibility from the sheep manager. There is a need to demonstrate that prescriptive allowances for space do indeed meet the described and tested needs of various classes and conditions of sheep. More space must be given if the behaviour of a group of sheep indicates the necessity. Pens can restrict the behaviours exhibited by sheep to maintain their physiological comfort in the short term. For example, sheep in pens must be allowed to lie down or stand up in the orientation they choose and at a time they choose. They must also be allowed to drink at the time they freely select. The standards for establishing whether sheep are not operating an unfettered choice of behaviour to maximize their own comfort must come from the systematic observation of behaviour and identification of stressors. Ethograms appropriate for the class, physiological status and genotype are a felt need for this purpose. They will be useful for accrediting or commissioning intensive production systems for adequate welfare and as key components in continuing quality assurance for adequate welfare.

Since sheep cannot escape from pens and seek shelter in a self-satisfying manner, they must not be exposed to conditions of heat, cold, noise, wetness or light and dark which encroach deeply into their physiological reserve for coping. In this connection, the microenvironment of sheep in pens and at pasture under the same nominal climatic conditions can be entirely different. Microclimate may vary markedly within the same set of pens. Good ventilation is crucial in intensively housed sheep and the full fleece which is an environmental advantage at pasture may become a handicap in pens. The question of noise is not clear at the moment but investigations should consider differences in the auditory range between sheep and people. Continuous wetness in pens is a distinct problem for freshly shorn sheep and is responsible for unacceptable levels of wound infection.

Sheep should not be exposed to physical damage from poorly designed flooring and fittings in sheds and feedlots. Any death or injury by misadventure or poor handling in intensively husbanded sheep is a clear indictment of poor husbandry. Types of slatted flooring which do not allow sheep to stand or lie in comfort or which become clogged with faeces are unacceptable. The question of lying and standing in comfort can be answered: (i) by general physical examination which can identify superficial lesions and problems with locomotion; and (ii) by observation of behaviour. Ethograms will have an application here. Sheep should display an appropriate pattern of lying and standing. Good design of housing for welfare can be more a matter of thought than expense. For example, it costs no more in temperate climates to position sheep housing such that it uses the winter sun for warmth and presents the best of shade in summer. It costs no more to align slatted flooring at right angles to the direction sheep move in passageways. This orientation provides sheep with the impression of solid flooring and prevents them baulking out of fear.

Other problems in pens relate to disease, the possibility of psychosocial stressors and nutrition. An increased exposure to certain pathogens may be partly responsible for different patterns of disease incidence in housed and lot-fed sheep. However, immunodeficiencies caused by stressors may ultimately determine whether or not disease occurs. For this reason, the incidence of disease can be an index of stress.

Disease

Two particular non-infectious disease problems which can occur in pens are copper toxicity and urinary calculi. Both diseases have a complex cause in nutrition and both raise welfare concern. Elaborate regimes are available for controlling copper toxicity. However, access to soil and its consumption by sheep is a useful preventive measure. Chemical analysis of calculi is necessary to determine their origin and to design effective control measures. The presence of urinary calculi should direct attention to the adequacy of the water supply. Clean water of adequate quality and uncontaminated by sheep droppings must be easily and continuously available to all sheep. Contamination of water with sheep, rat or bird droppings and algae indicates poor pen and water trough design, overcrowding, slapdash husbandry and compromised welfare. The same indictment of poor husbandry applies if diseases such as blowfly-strike and posthitis ('sheath-rot') occur in intensively housed sheep.

Nutrition

Nutrition presents two issues for the welfare of intensively husbanded sheep. Firstly, the ration must be adequate in amount and composition to provide sheep with their physiological needs. Secondly, the ration must provide sheep with their behavioural needs. In this regard, little is known about hunger in sheep. The general phenomenon of hunger in mammals has at least two aspects. Primary hunger stems from deprivation of food and has a distinctly thermodynamic basis. Specific hungers relate to particular dietary components such as salt. The situation, however, may be more complex for ruminants such as sheep. It is unclear whether cud chewing operates independently from hunger and from feeding and foraging behaviour. Cud chewing or rumination may be a continuously present behavioural need which must be satisfied irrespective of nutrition. Pelleted rations have not been accredited for their provision of adequate cud for chewing. In a similar vein, daily feeding of fibre to intensively husbanded sheep may be both good nutritional and behavioural practice in that it provides almost continuous access to cud.

The failure to provide an adequate ration and to distribute it sufficiently to all individuals remains the greatest welfare problem of penned sheep. The problem is put into clearest perspective if mortalities in pens are higher than would be expected in sheep at pasture and with standard measures for disease control in place. Mortalities are likely to result from inadequate nutrition itself, i.e. starvation, and from low resistance to intercurrent disease caused by poor nutrition. In both instances, unacceptable distress will be present in more sheep than those which died. Aspects of this nutritional problem may be complex. However, if trough space is inadequate and if the amount of food provided is determined in a literal-minded fashion and is insufficient for individual sheep, it seems pointless to analyse the complexities. In some instances, the amount of feed given each day to a pen of sheep has been determined by multiplying the maintenance ration for individual sheep by the number of sheep in a pen. This approach takes no account of dominance behaviour in sheep nor of differences in the maintenance requirements for individual sheep. The problem is aggravated when there is insufficient trough space and more so when groups contain mixtures of large and small sheep.

Psychosocial stressors

The question of psychosocial stressors in penned sheep is problematical and the following evaluation does not have experimental backing. Nevertheless, the question should be explored, and the benefit of doubt should go to the sheep. Psychosocial stressors in pens relate to bullying and dominance behaviour and the absence of routes of escape for those sheep which rank lower in order of dominance. Inadequate space or a poor mix of sheep or both can contribute to the problem. Sheep from different sources and unfamiliar with one another require a period of adjustment. The process of socialization can consume behavioural resources. Sheep which differ markedly in size should not be penned together. Breeds of sheep with markedly different behaviour should not be penned together; for example, Merinos and Dorset Horns. Polled and horned sheep should not be penned together. In addition, horned sheep require more space. Rams present particular problems because of their more vigorous agonistic encounters with one another.

Wool-biting

Wool-biting is a particular problem of intensively housed sheep and is virtually non-existent in sheep at pasture. The cause of this abnormal behaviour directed towards another animal is complex. It is not treated as a stereotypy by Fraser and Broom (1990). Wool-biting is demonstrably related to penning because the behaviour disappears when sheep are returned to pasture. Indeed, release from enclosure is the only satisfactory treatment. Under these circumstances wool-bitten sheep are allowed room for escape and biters lose the habit. Provision of fibrous food appears not to reverse the practice once it is in place and wool-biting occurs in penned sheep fed long-stranded hay and straw. Dominance behaviour is involved and lower-ranking sheep are the usual target. Skin lesions occur when wool-biting goes unchecked. There is a clear tendency for the behaviour to escalate continuously. Wool-biting invalidates experiments made with helminth parasites in pens. Distinctly higher burdens with the stomach worm, *Haemonchus contortus*, occur in wool-bitten sheep and the range of worm burdens increases in groups of sheep where wool-biting is evident. Stress-induced immunodeficiency is probably involved.

Stereotypic behaviour

Interestingly, stereotypic behaviour has not been described in great detail for the sheep. Stereotypies are abnormal behaviours. The term refers to a 'repeated, relatively invariate sequence of movements which has no obvious purpose' (Fraser and Broom, 1990). Examples are wind-sucking in horses and sham-chewing in sows. Stereotypies are argued to relate to a need for stimulation from the environment and to act as a compensating mechanism for maintaining behavioural homeostasis. In short, stereotypies may point to existence of deficiencies in a confined environment which have implications for welfare. It has been argued that stereotypies are not associated with suffering in the animals which express them (Dantzer, 1986). However, this line of thought raises the possibility of suffering in animals which do not adapt by expressing stereotypies. Accordingly, attention is redirected to environmental deficiencies and their welfare implications. Stereotypies may not be a striking feature of the behaviour of the sheep. They may follow the pattern of other behaviours in this species and be expressed in a subtle or even cryptic manner when compared with stereotypies in other domestic animals. The danger is that stereotypies may have been missed in the sheep. They may only be detected if deliberately and systematically investigated. For example, they have been described in studies made of the behaviour transition occurring when sheep are moved from pasture and placed in pens (Done-Currie *et al.*, 1984).

Welfare issues associated with extensive systems of sheep husbandry

The specific welfare problems of grazing sheep are different from those of penned sheep. At pasture, there is usually enough room to escape psycho-social stressors and the issues for welfare relate to factors operating cyclically and not continuously. These factors are climate, nutrition and disease. They are interrelated, interdependent and they arise from the environment. For this reason, grazing systems provide a very close match between the requirements for adequate welfare and those for production performance. Concerns about stewardship for the wider environment, the ecological sustainability of animal-based agriculture and the welfare of sheep overlap and merge in grazing systems. Stocking densities which are too high and which result in overgrazing lead to a continuing downward spiral of under-nutrition, damage to pastures and soil, further undernutrition and ecological non-sustainability.

The description of welfare as the state of an individual as regards its attempts to cope with its environment (Broom, 1986) is particularly relevant to sheep managed under extensive grazing systems. The match between the genotype of sheep and a given environment is vitally important to welfare and the wide genetic polymorphism of sheep has allowed adaptations to a wide range of climatic and nutritional conditions. Sheep have been placed into three major groups according to their environments and their implied adaptations (Ryder, 1983). First is the southern desert group located in the Sahara and on the hot plains of India and in south-east Asia. Sheep in this group have short kempy hair coats and long bodies and extremities. Then comes the northern desert group with more compact bodies and a coarse-woolled fleece suitable for carpets. This group includes fat-tailed sheep such as the Awassi. Finally, there is the temperate group consisting of sheep with compact bodies and coarse to fine wools. It includes European breeds such as the Merino.

The geographical distribution of sheep breeds is a fact so obvious and familiar to sheep husbandry that it may be overlooked as being an important component of welfare. Sheep suited to one environment may be ill-suited to another and their relocation can be a source of suffering. Illustrations of this point can be proposed throughout the world. For example, Merino sheep can suffer from diseases of the skin and fleece in areas of high rainfall.

The specific welfare problems of extensively grazed sheep relate to disease, climate and nutrition. Parasitic disease is a pressing problem for sheep and is well treated in specialty textbooks. Climate and nutrition are intimately related in grazing situations. Rainfall determines the amount of available grazing. Management of this uncertainty throughout most of the world becomes an issue for welfare as well as for production. Good welfare practice demands that sheep flocks be managed to avoid feed shortages or to make up the shortfall with supplements. Computer-based decision support systems can be valuable here. Unfortunately, this approach may be possible only in rich developed countries. In poorer countries sheep must suffer the vagaries of the environment along with their human stewards.

Some of the problems of confinement relate to sheep under grazing systems as well as to intensively housed sheep. For example, sheep restricted to particular paddocks may not have access to suitable shelter especially in cold, wet and windy conditions. Newborn lambs are particularly vulnerable to such conditions and the provision of shelter can increase the yield of live lambs (Lynch and Alexander, 1977). Shelter in this situation gives both a production and welfare benefit and repeats the theme that production performance and good welfare are intimately related.

Welfare Issues Associated with Husbandry Operations

Husbandry operations can be simple and relatively non-invasive or can involve surgery of a minor nature. Both categories will be considered in this section. The first category includes general handling, dipping for external parasites, drenching for internal parasites and so on. The second includes invasive husbandry operations such as castration and tail-docking. These were described as mutilations in the report of the Brambell Committee and the description has persisted in official documents and elsewhere. While these operations can confer welfare benefits over the long term, they raise welfare concern in the short term because they involve surgery and tissue damage. Thus, general husbandry operations such as simple handling may entail the motivational states of fear, anxiety and frustration. Husbandry operations involving surgery may include these same motivational states plus physical pain. The approach towards welfare for the latter set of husbandry operations could make use of the three Rs which have been described for animal experimentation (Russell and Burch, 1959). These are replacement which refers to the search for a more acceptable means to the same end, refinement which refers to ways in which a given procedure may be alleviated, and reduction which refers to ways in which it might be possible to limit the number of times an unpleasant operation need be performed.

As stated previously, physical pain occurs in two forms. First pain relates to the process of tissue damage during surgery and second or chronic pain relates to the consequences of tissue damage. Post-operative pain fits into the latter category. These two aspects of pain and general pain-related behaviour mentioned earlier are pertinent to welfare assessments of husbandry operations. Pain and production performance are connected in this matter. Pain has been described as a 'need-state' (Wall, 1979) involving private and unpleasant feeling-tones plus observable physiological consequences. Some of the physiological consequences of pain have implications for production. Processes such as body growth and wool growth can be depressed by the catabolic action of hormones such as the corticosteroids which are released as part of the overall pain response. In addition, corticosteroids are immunosuppressive and the pain state can be accompanied by reduced resistance to disease, especially parasitic disease in sheep.

Castration and tail-docking

There are legal requirements in many countries for the castration and tail-docking of sheep. These are based on obvious anatomical and physiological considerations and on the need to cope with parasites such as blowflies. Legal requirements relate to the acceptability of various forms of castration at various ages and the age at which anaesthesia becomes mandatory. However, there is now a need to establish the acceptability of various methods for castration and tail-docking on a more defensible basis. A systematic and extremely promising start has been made, but more work is called for in this difficult area to clarify the picture regarding first and second pain over a longer time frame. Assessment could include examination of healing times and the effect of a given procedure and associated handling on the integrated physiological responses of animals. It is known that elastrator rings are associated with tetanus in sheep in Australia and the use of elastrator rings for tail-docking is contraindicated while the sheep blowfly is active. Here, the flaccid tail still attached to the lamb creates a desirable situation for flystrike.

Behavioural and physiological comparisons of castration and tail-docking with elastrator rings or the knife have been made by two groups (Shutt *et al.*, 1988; Mellor and Murray, 1989). If behaviour is taken as a guide, these two methods appear to be associated with different qualities of pain in the short term. Elastrator rings cause pain associated with writhing. The knife causes pain perhaps associated with immobility.

Blowfly strike and the Mules operation

Blowfly strike is a major disease and welfare problem of sheep worldwide. The issue is given an added dimension in Australia where the Mules operation together with the docking of tails to an appropriate length is employed as a standard measure for the control of flystrike, especially in Merino sheep. The Mules operation consists of the surgical removal of folds of skin around the tail of sheep to make the site less desirable for oviposition by blowflies.

There is a welfare dilemma here. On the one hand, mulesing of sheep for control of blowfly strike is undergoing scrutiny because of its invasiveness and because it is performed without anaesthesia. On the other hand, the Mules operation plus correct tail-docking can provide effective and enduring protection against flystrike. It may be indispensable to strategies to control sheep blowfly and may become more important as blowfly resistance to insecticides becomes more prevalent. Indeed, the failure to carry out the Mules operation could be interpreted as neglect. The use of mulesing is based upon judgements about the balance of benefits and harm. Thus, the pain and suffering associated with flystrike is assessed as being far greater than the pain and suffering associated with mulesing. As to the suffering caused by flystrike, it could be enormous if the extent of tissue damage and the associated release of neuro- and immunomodulators is considered. Death from flystrike appears to come from peripheral circulatory failure or shock.

This simple analysis, however, does not pose the question whether mulesing need be a component of integrated blowfly control programmes in all situations. Moreover, it does not consider the detrimental effects of mulesing in some instances. These relate to productivity as well as to welfare. In this connection, mulesing must be carried out correctly. The operation of mulesing must not be allowed to spread infectious diseases. Examples are eperythrozoonosis and bacterial septicaemia and polyarthritis. It is not possible to handle sheep for many weeks after mulesing and this may interfere with programmes for controlling other diseases such as footrot and parasitism.

It is noteworthy that welfare considerations about mulesing have centred on the operation itself and very little on the post-operative period. Mulesing is associated with high plasma concentrations of the opioid peptide, β-endorphin (Shutt *et al.*, 1987). Without doubt, the presence of this peptide implies an effect on physiology and behaviour. It may also indicate some analgesia but this connection must be demonstrated.

The three Rs of reduction, refinement and replacement which are used in animal experimentation (Russell and Burch, 1959) can help in improving the welfare aspects of mulesing. For instance, wound dressings may enhance wound healing (Levot *et al.*, 1989) and in this way reduce the time during which post-operative pain may occur. Wound dressings may even reduce the intensity of post-operative pain as determined by a suite of behavioural and physiological observations.

General handling, dipping and shearing

Handling of animals can induce fear, anxiety and frustration and can, for this reason, raise welfare concern. The distinction between arousal and stress is important in this regard. Much of the response to handling occurs within the physiological and behavioural reserve of domestic animals such as the sheep and may not necessarily lead to stress or, as it has been put in the present chapter, to diseases of adaptation. A systematic evaluation of the physiological and behavioural responses of sheep to a series of handling procedures including shearing has been commenced (Hargreaves and Hutson, 1990). One finding was an increase in the plasma concentration of cortisol within 20 minutes of a single non-invasive procedure; running sheep down a drafting race.

Poor welfare practice can occur during the routine practice of mustering sheep in the field when this is carried out with motor vehicles. The problem is the tendency to move sheep too quickly and to push them to the limit of their exercise tolerance. This misjudgement does not arise when sheep are mustered on foot or on horseback. The remedy is to take extra pains in observing sheep when they are being mustered by motor vehicle.

Welfare Issues Associated with Transportation

Transportation by road, rail, sea or air is responsible for a range of infringements against good welfare. These infringements increase with the duration of transport and the distance travelled. They are aggravated under harsh climatic conditions. The welfare problems associated with transport have straightforward causes. They are associated with overcrowding, physical trauma, climatic stress, change of diet, hunger and thirst. Observations of bruising, carcase damage and meat quality at slaughterhouses can be used in quality control for the welfare adequacy of transport systems and can draw attention to both specific and more general problems. More specific observations could include the measurement of dehydration and glycogen stores. Records of elevated plasma concentrations of adrenal corticoid hormones in transported sheep appear in the scientific literature.

There is an economic scenario which has unfortunate ramifications for the welfare of sheep. Under this scenario, losses from death and disease during transportation are sometimes treated as a business risk which can be dealt with by the usual insurance procedures for risk management rather than by rectifying physical problems. The cost of mortalities is externalized to the wider community and no action is taken in favour of the welfare of sheep.

Welfare Issues Associated with Slaughter

Slaughter is likely to remain a contentious welfare issue despite the use of demonstrably humane methods. Arguments about humane slaughter can be confused with the more general argument about the prerogative for people to kill animals for food. There are inherent aesthetic qualms about the slaughter of animals. These are augmented by the projection of a universal human concern into the situation. Do animals have concepts of death and are the uncertainties a cause of suffering for them as they can be for people? Finally, differences in cultural and religious outlooks create different requirements for slaughter and these are a source of potential conflict. Two ideas are relevant. First, humane care of animals is common to all the major religions of the world. Secondly, scientific analysis can provide some common ground for the resolution of this conflict. It must be recognized, however, that scientific research will not provide the total solution to religious and cultural differences about the methods used to slaughter animals for food.

The issues associated with slaughter are common to all domestic animals (UFAW, 1987; Blackmore and Delany, 1988) but there are significant species differences in the response to various methods. Welfare problems associated with slaughter occur at two points. First is during the period leading up to slaughter including the time of lairage. The issues here are general and have been dealt with elsewhere in this chapter. The second point for welfare concern is during the slaughter process itself.

It is becoming clearer that quality control procedures for the continuing effectiveness of any given method for slaughter may be as important to welfare as the method itself (Blackmore and Delany, 1988). This quality control theme can be used to analyse current methods used to slaughter sheep. The three major methods are (i) exsanguination by severing the major blood vessels in the neck without previous stunning, (ii) exsanguination the same way but preceded by electrical or percussive stunning, and (iii) penetrative stunning which causes death itself and which is followed by exsanguination. Variations on electrical stunning are aimed at producing either insensibility alone or combining insensibility with cardiac arrest.

Some comments are required before proceeding to analysis of the various methods for slaughtering sheep from the viewpoint of welfare.

1. Sheep differ from cattle anatomically. The carotid arteries of the sheep are the major source of blood to the brain. Paravertebral arteries are virtually irrelevant. This means that exsanguination *via* severed blood vessels in the neck will result in a virtually complete interruption of the blood supply to the brain.
2. Sheep lose brain function and by this criterion are dead within an extremely short period (15 seconds) after the commencement of exsanguination (see UFAW, 1987). Analgesia and anaesthesia could be expected somewhat earlier. It is during this short period that first pain related to the threat and process of tissue damage could be registered in the brain and provide a subjective mental experience.
3. Stunning was recommended as a procedure to enhance the humanity of slaughter methods well before there were methods for accrediting the effectiveness of the unconsciousness produced. The connection between electrical stunning and its welfare rationale in an epileptiform state plus an associated analgesia or anaesthesia is relatively recent (see UFAW, 1987).
4. The criteria used to infer loss of sensibility, unconsciousness and death have often been flawed. The stages of anaesthesia recognized for surgical purposes when a general anaesthetic agent is used may not be appropriate for understanding the unconsciousness produced by brain anoxia during slaughter. The brain function which supports consciousness is easily damaged by anoxia whereas the reflex muscular reactions associated with

breathing are extremely resistant to the same conditions. Other criteria such as visual evoked potentials (see UFAW, 1987) have been extrapolated directly from human medicine. Visual evoked potentials measure mid-brain function and are used to determine whether the continuation of life-support systems in intensive care is a reasonable proposition. Finally, the mechanism underlying some of the body responses during exsanguination runs against common perceptions. Limb paddling and flailing after the neck vessels of sheep have been severed and in the presence of an intact spinal cord can be considered to reflect a lack of descending control from the brain on these activities and, accordingly, the absence of this brain function.

Given the considerations just listed, the following analysis of the welfare aspects of various slaughter methods can be proposed.

1. Exsanguination by severing the major blood vessels in the neck of sheep is an irreversible process leading to rapid and progressive loss of consciousness. It will cause certain death within a very short period. It is in the short period between severing the neck vessels, 'sticking', and loss of consciousness that the possibility of physical pain and attendant suffering can be realized.

2. Percussive or electrical stunning before 'sticking' must not impose suffering additional to or equal to the suffering possible in the short period after the neck vessels have been severed without stunning. To do this, stunning must be effective in producing immediate unconsciousness in all sheep which must remain until loss of blood prevents a return to consciousness. Stunning may confer no welfare benefits and may be a welfare hazard if it is not performed effectively. In such cases, stunning and sticking will be less humane than sticking alone. Stunning must be accompanied by rigid guidelines for and diligent execution of quality control procedures.

Conclusions

This chapter has followed the present trend of attempting to provide a scientifically defensible set of criteria for use in the care and welfare of sheep. It is clear that these criteria and the eventual delivery of adequate welfare depend heavily upon an enhanced understanding of behaviour. Given the extreme adaptability of sheep as a species, the modification of husbandry practices to facilitate welfare should not be difficult.

The connection between good welfare practice and production performance has been a major theme and for two reasons. Firstly, the connection does indeed exist. Secondly, concentration in the past on those situations where aspirations for welfare and production are seen to conflict has had a diversionary, unintentional and unfortunate impact on public debate. It has inhibited the introduction of practices which can benefit both welfare and production. The institution of husbandry practices which improve both welfare and production must be seen as an important advance providing the necessary starting point for addressing the residue of welfare problems and certainly benefiting the production process.

Further Reading

Dawkins, M.S. (1980) *Animal Suffering: The Science of Animal Welfare.* Chapman and Hall, London.
Dawkins, M.S. (1990) From an animal's point of view: motivation, fitness and animal welfare. *Behavioural and Brain Sciences* 13, 1–61.
Ewbank, R. (1988) Animal welfare. In: *UFAW Handbook: Management and Welfare of Farm Animals.* Ballière Tindall, London, pp. 1–12.
Fraser, A.F. and Broom, D.M. (1990) *Farm Animal Behaviour and Welfare.* Ballière Tindall, London.

References

Adams, D.B. (1988) An approach to pain in research animals. *Alternatives to Laboratory Animals* 16, 145–54.

Adams, R.D. and Martin, J.B. (1983) Pain. In: Petersdorf, R.G., Adams, R.D., Braunwald, K., Isselbacher, K.J., Martin, J.B. and Wilson, J.D. (eds), *Harrison's Principles of Internal Medicine*, 10th edn. McGraw-Hill, New York, pp. 7–13.

Ader, R., Grota, L.J. and Cohen, N. (1987) Conditioning phenomena and immune function. *Annals of the New York Academy of Sciences* 496, 532–44.

Alexander, G. (1984) Constraints to lambs' survival. In: Lindsay, D.R. and Pearce, D.T. (eds), *Reproduction in Sheep*. Australian Academy of Science and the Australian Wool Corporation, Canberra, pp. 199–209.

Alexander, G. and Shillito, E. (1978) Maternal responses in Merino ewes to artificially coloured lambs. *Applied Animal Ethology* 4, 141–52.

Alexander, G., Lynch, J.J., Mottershead, B.E. and Donnelly, J.B. (1980) Reduction in lamb mortality by means of grass windbreaks: results of a five-year study. *Proceedings of the Australian Society of Animal Production* 13, 329–32.

Alexander, G., Stevens, D., Kilgour, R., de Langen, H., Mottershead, B.E. and Lynch, J.J. (1983) Separation of ewes from twin lambs: incidence in several sheep breeds. *Applied Animal Ethology* 10, 301–17.

Alexander, G., Poindron, P., Le Neindre, P., Stevens, D., Levy, F., and Bradley, L. (1986) Importance of the first hour post-partum for exclusive maternal bonding in sheep. *Applied Animal Behaviour Science* 16, 295–300.

Alexander, G., Stevens, D. and Bradley, L.R. (1988) Maternal behaviour in ewes following caesarian section. *Applied Animal Behaviour Science* 19, 273–7.

Alexander, G., Stevens, D. and Bradley, L.R. (1989) Fostering in sheep: an exploratory comparison of several approaches. *Australian Journal of Experimental Agriculture* 29, 509–12.

Alexander, G., Stevens, D., Bradley, L.R. and Barwick, S.A. (1990) Maternal behaviour in Border Leicester, Glen Vale (Border Leicester derived) and Merino sheep. *Australian Journal of Experimental Agriculture* 30, 27–38.

211

Allden, W.G. and Whittaker, I.A.McD. (1970) The determinants of herbage intake by grazing sheep: the interrelationships of factors influencing herbage intake and availability. *Australian Journal of Agricultural Research* 21, 755–66.

Allison, A.J. and Davis, G.H. (1976a) Studies of mating behaviour and fertility of Merino ewes. I. Effect of number of ewes joined per ram, age of ewe and paddock size. *New Zealand Journal of Experimental Agriculture* 4, 259–67.

Allison, A.J. and Davis, G.H. (1976b) Studies of mating behaviour and fertility of Merino ewes. II. Effects of age of ewe, live weight and paddock size on duration of oestrus and ram seeking activity. *New Zealand Journal of Experimental Agriculture* 4, 269–75.

American Veterinary Medical Association (1987) Colloquium on recognition and alleviation of animal pain and distress. *Journal of the American Veterinary Medical Association* 161, 1186–296.

Anderson, D.M., Hulet, C.V., Shupe, W.L., Smith, J.N. and Murray, L.W. (1988) Response of bonded and non-bonded sheep to the approach of a trained Border Collie. *Applied Animal Behaviour Science* 21, 251–7.

Anonymous (1991) Crops and livestock. In: *Britannica: Book of the Year.* Encyclopedia Britannica, Chicago, pp. 788–93.

Arnold, G.W. (1981) Grazing behaviour. In: Morley, F.H.W. (ed.), *Grazing Animals: World Animal Science.* Elsevier, Amsterdam, pp. 79–104.

Arnold, G.W. (1985) Association and social behaviour. In: Fraser, A.F. (ed.), *Ethology of Farm Animals.* Elsevier, Amsterdam, pp. 233–46.

Arnold, G.W. (1985) Parturient behaviour. In: Fraser, A.F. (ed.), *Ethology of Farm Animals.* Elsevier, Amsterdam, pp. 335–47.

Arnold, G.W. and Dudzinski, M.L. (1978) *Ethology of Free Ranging Domestic Animals.* Elsevier Scientific Publishers, Amsterdam.

Arnold, G.W. and Maller, R.A. (1974) Some aspects of the social competition between sheep for supplementary feed. *Animal Production* 19, 309–19.

Arnold, G.W. and Morgan, P.D. (1975) Behaviour of the ewe and lamb at lambing and its relationship to lamb mortality. *Applied Animal Ethology* 2, 25–46.

Arnold, G.W. and Pahl, P.J. (1967) Subgrouping in sheep flocks. *Proceedings of the Ecological Society of Australia* 2, 183–9.

Arnold, G.W., Wallace, S.R. and Maller, R.A. (1979) Some factors involved in natural weaning processes in sheep. *Applied Animal Ethology* 5, 43–50.

Arnold, G.W., Campbell, N.A. and Pahl, P.J. (1981) The effect of age and breed on diet selection by sheep. *Land Resources Management Technical Paper* No. 9, 11, CSIRO, Australia.

Arnold, G.W., Wallace, S.R. and Rea, W.A. (1981) Associations between individuals and home-range behaviour in natural flocks of three breeds of domestic sheep. *Applied Animal Ethology* 7, 239–57.

Baile, C.A. and McLaughlin, C.L. (1987) Mechanisms controlling feed intake in ruminants: a review. *Journal of Animal Science* 64, 915–22.

Baldock, N.M., Sibly, R.M. and Penning, P.D. (1988) Behaviour and seasonal variation in heart rate in domestic sheep (*Ovis aries*). *Animal Behaviour* 36, 35–43.

Baldwin, B.A. and Start, I.B. (1981) Sensory reinforcement and illumination preference in sheep and calves. *Proceedings of Royal Society of London* 211, 513–26.

Banks, E. (1964) Some aspects of sexual behaviour in domestic sheep (*Ovis aries*). *Behaviour* 23, 249–79.

Barcroft, J. and Barron, D.H. (1937) Movements in midfoetal life in the sheep embryo. *Journal of Physiology* 91, 329–51.

Barcroft, J. and Barron, D.H. (1939) The development of behaviour in foetal sheep. *Journal of Comparative Neurology* 70, 477–502.

Bareham, J.R. (1976) The behaviour of lambs on the first day after birth. *British Veterinary Journal* 132, 152–61.

Barthram, G.T. (1980) Sward structure and the depth of the grazed horizon. Proceedings of the British Grassland Society Winter Meeting 1980. *Grass and Forage Science* 36, 130–1.

Baskin, L.M. (1974) Management of ungulate herds in relation to domestication. In: Geist, V. and Walther, F. (eds), *The Behaviour of Ungulates in Relation to Management.* IUCN, Morges, Switzerland, pp. 530–41.

Baskin, L.M. (1975) Distribution of animals on pasture as a function of group behaviour. *Selskokhoz Biology* 10, 407–11.

Bateman, A., Singh, A., Kral, T. and Solomon, S. (1989) The immune–hypothalamic–pituitary–adrenal axis. *Endocrine Reviews* 10, 92–113.

Bentham, J. (1789) In: Harrison, W. (ed.) (1967) *An Introduction to the Principles of Morals and Legislation.* Basil Blackwell, Oxford.

Berger, J. (1979) Weaning conflict in Desert and Mountain Bighorn sheep (*Ovis canadensis*): an ecological interpretation. *Zeitschrift Tierpsychologie* 50, 188–200.

Birch, L.C. (1990) *On Purpose.* New South Wales University Press, Sydney, Australia.

Blackmore, D.W. and Delany, M.W. (1988) *Slaughter of Stock: a Practical Review and Guide.* Veterinary Continuing Education Publication, Massey University, Palmerston North, New Zealand.

Blalock, J.E. (1988) Production of neuroendocrine peptide hormones by the immune system. *Progress in Allergy* 43, 1–13.

Blissett, M.J., Boland, K.P. and Cottrell, D.F. (1990) Discrimination between odours of fresh oestrous and non-oestrous ewe urine by rams. *Applied Animal Behaviour Science* 25, 51–9.

Boyd, J.M., Doney, J.M., Gunn, R.G. and Jewell, P.A. (1964) The Soay sheep of the island of Hirta, St. Kilda: a feral population. *Proceedings of the Zoological Society, London* 142, 129–63.

Bradley, R.M. and Mistretta, C.M. (1975) Fetal sensory receptors. *Physiological Reviews* 55, 352–75.

Brain, R. (1962) Presidential address. In: Keele, C.A. and Smith, R. (eds), *The Assessment of Pain in Man and Animals.* Universities Federation for Animal Welfare, E. and S. Livingstone, Edinburgh, pp. 3–11.

Brambell, F.H.W. (1965) *Report of the Technical Committee to Enquire into the Welfare of Animals Kept Under Intensive Livestock Husbandry Systems.* Her Majesty's Stationery Office, London.

Bremner, K.J., Braggins, J.B. and Kilgour, R. (1980) Training sheep as leaders in abattoirs and farm sheep yards. *Proceedings of the New Zealand Society of Animal Production* 40, 111–16.

Broom, D.M. (1986) Indicators of poor welfare. *British Veterinary Journal* 142, 524–6.

Broom, D.M. (1989) Animal welfare. In: Grunsell, C.S.G., Raw, M.E. and Hill,

F.W.G. (eds) *The Veterinary Annual,* Vol. 29. Wright, London, pp. 9–14.

Broom, D.M. and Arnold, G.W. (1986) Selection by grazing sheep of pasture plants at low herbage availability and responses of the plants to grazing. *Australian Journal of Agricultural Research* 37, 527–38.

Burritt, E.A. and Provenza, F.D. (1989) Food aversion learning: ability of lanbs to distinguish safe from harmful foods. *Journal of Animal Science* 67, 1732–9.

Burritt, E.A. and Provenza, F.D. (1990) Food aversion learning in sheep: persistence of conditioned taste aversions to palatable shrubs (*Cercocarpus montanus* and *Amelanchier alnifolia*). *Journal of Animal Science* 68, 1003–7.

Burritt, E.A. and Provenza, F.D. (1991) Lambs form preferences for non-nutritive flavors paired with calories. *Journal of Animal Science* (in press).

Casteilla, L., Orgeur, P. and Signoret, J.P. (1987) Effects of rearing conditions on sexual performance in the ram: practical use. *Applied Animal Behaviour Science* 19, 111–18.

Chapple, R.S. and Lynch, J.J. (1986) Behavioral factors modifying acceptance of supplementary foods by sheep. *Research Development in Agriculture* 3, 113–20.

Chapple, R.S., Wodzicka-Tomaszewska, M. and Lynch, J.J. (1987) The learning behaviour of sheep when introduced to wheat. II. Social transmission of wheat feeding and the role of the senses. *Applied Animal Behaviour Science* 18, 163–72.

Craig, J.V. (1986) Measuring social behaviour: social dominance. *Journal of Animal Science* 62, 1120–9.

Crofton, H.D. (1958) Nematode parasite populations in sheep on lowland farms. VI. Sheep behaviour and nematode infections. *Parasitology* 48, 251–60.

CSIRO (1976) Understanding sheep behaviour. *Rural Research in CSIRO* 93, 4–10.

CSIRO (1982) Back to nature sheep raising. *Rural Research in CSIRO* 114, 20–2.

CSIRO (1985) Reducing lamb losses by fostering. *Rural Research in CSIRO* 128.

Cumming, R.B. (1989) Exploiting the appetites of the hen. In: *Australian Poultry Science Symposium 1989.* Poultry Research Foundation and World Poultry Science Association, Sydney, pp. 38–44.

Curll, M.L., Wilkins, R.J., Snaydon, R.W. and Shanmugalingham, V.S. (1985) The effects of stocking rate and nitrogen fertilizer on a perennial ryegrass-white clover sward. 1. Sward and sheep performance. *Grass and Forage Science* 40, 129–40.

Dantzer, R. (1986) Behavioural, physiological and functional aspects of stereotyped behaviour: a review and reinterpretation. *Journal of Animal Science* 62, 1776–86.

Davidson, R.J. (1986) Emotion: satiation and starvation effects on. In: Harré, R. and Lamb, R., *The Dictionary of Ethology and Animal Learning.* Basil Blackwell, Oxford, p. 6.

Davis, W. (1938) Summer activity of mountain sheep on Washburn, Yellowstone National Park. *Journal of Mammology* 19, 88–94.

Dawkins, M.S. (1980) *Animal Suffering: the Science of Animal Welfare.* Chapman & Hall, London.

Dawkins, M.S. (1990) From an animal's point of view: motivation, fitness and animal welfare. *Behavioural and Brain Sciences* 13, 1–61.

Denholm, L.J. and Vizard, A.L. (1986) Trimming the incisor teeth of sheep: another view. *Veterinary Record* 119, 182–4.

Done-Currie, J.R., Hecker, J.F. and Wodzicka-Tomaszewska, M. (1984) Behaviour of sheep transferred from pasture to an animal house. *Applied Animal Behaviour Science* 12, 121–30.

du Toit, J.T., Provenza, F.D. and Nastis, A.S. (1991) Conditioned food aversions: how sick must a ruminant get before it detects toxicity in foods? *Applied Animal Behaviour Science* 30, 35–46.

Eccles, T.R. and Shackleton, D.M. (1986) Correlates and consequences of social status in female bighorn sheep. *Animal Behaviour* 34, 1392–401.

Ewbank, R. (1985) The behavioural needs of farm and laboratory animals. In: Marsh, N. and Haywood, S. (eds), *Animal Experimentation: Improvements and Alternatives.* Fund for the Replacement of Animals in Medical Experiments, Nottingham, pp. 31–5.

Ewbank, R. (1988) Animal welfare. In: *UFAW Handbook: Management and Welfare of Farm Animals.* Ballière Tindall, London, pp. 1–12.

Fell, L.R. and Shutt, D,.A. (1989) Behavioural and hormonal responses to acute surgical stress in sheep. *Applied Animal Behaviour Science* 22, 283–94.

Fell, L.R., Lynch, J.J., Adams, D.B., Hinch, G.N. and Munro, R.K. (1991) Behavioural and physiological effects in sheep of a chronic stressor and a parasite challenge. *Australian Journal for Agricultural Research* 42, 1135–46.

Felten, S.Y., Felten, D.L., Bellinger, D.L., Carison, S.L., Ackerman, K.D., Madden, K.S., Olsschowka, J.A. and Livnat, S. (1988) Noradrenergic sympathetic innervation of lymphoid organs. *Progress in Allergy* 43, 14–67.

Fletcher, I.C. and Lindsay, D.R. (1968) Sensory involvement in the mating behaviour of domestic sheep. *Animal Behaviour* 16, 410–14.

Flores, E.R., Provenza, F.D. and Balph, D.F. (1989a) Role of experience in the development of foraging skills of lambs browsing the shrub serviceberry. *Applied Animal Behaviour Science* 23, 271–8.

Flores, E.R., Provenza, F.D. and Balph, D.F. (1989b) Relationship between plant maturity and foraging experience of lambs grazing hycrest crested wheatgrass. *Applied Animal Behaviour Science* 23, 279–84.

Forbes, T.D.A. and Beattie, M.M. (1987) Comparative studies of ingestive behaviour and diet composition in oesophageal-fistulated and non-fistulated cows and sheep. *Grass and Forage Science* 42, 79–84.

Ford, J.J. and D'Occhio, M.J. (1989) Differentiation of sexual behaviour in cattle, sheep and swine. *Journal of Animal Science* 67, 1816–23.

Fowler, D.G. (1984) Reproductive behaviour of rams. In: Lindsay, D.R. and Pearce, D.T. (eds), *Reproduction in Sheep,* Australian Academy of Science and The Australian Wool Corporation, Canberra, pp. 39–46.

Fowler, D.G. and Jenkins, L.D. (1976) The effects of dominance and infertility of rams on reproductive performance. *Applied Animal Ethology* 2, 327–37.

Fox, M.W. (1986) *Laboratory Animal Husbandry.* State University of New York Press, Albany, NY.

Franklin, J.R. and Hutson, G.D. (1982a) Experiments on attracting sheep to move along a laneway. I. Olfactory stimuli. *Applied Animal Ethology* 8, 439–46.

Franklin, J.R. and Hutson, G.D. (1982b) Experiments on attracting sheep to move along a laneway. II. Auditory stimuli. *Applied Animal Ethology* 8, 447–56.

Franklin, J.R. and Hutson, G.D. (1982c) Experiments on attracting sheep to move along a laneway. III. Visual stimuli. *Applied Animal Ethology* 8, 457–78.

Fraser, A.F. (1985) *Ethology of Farm Animals.* Elsevier, Oxford.

Fraser, A.F. (1989) The phenomenon of pandiculation in the kinetic behaviour of the sheep fetus. *Applied Animal Behaviour Science* 24, 169–82.

Fraser, A.F. and Broom, D.M. (1990) *Farm Animal Behaviour and Welfare.* Ballière Tindall, London.

Fraser, A.H.H. (1926) Chain instincts in lambing sheep. *Journal of Psychology* 16, 311–13.

Freedland, W.J. and Janzen, D.H. (1974) Strategies in herbivory by mammals: the role of plant secondary compounds. *American Naturalist* 108, 269–89.

Geist, V. (1966) The evolution of horn-like organs. *Behaviour* 27, 175–214.

Geist, V. (1971) *Mountain Sheep – a Study in Behaviour and Evolution.* University of Chicago Press, Chicago.

Geist, V. (1984) Goat antelopes. In: Macdonald, D. (ed.), *Hoofed Mammals.* Torstar Books, New York, pp. 144–51.

Gibson, T.E. (ed.), (1988) *Animal Disease – A Welfare Problem?* British Veterinary Association Welfare Foundation, London.

Gillingham, M.P. and Bunnell, F.L. (1989) Effects of learning on food selection and searching behaviour of deer. *Canadian Journal of Zoology* 67, 24–32.

Gluesing, E.A. and Balph, D.F. (1980) An aspect of feeding behaviour and its importance to grazing systems. *Journal of Range Management* 33, 426–7.

Goddard, M.E. (1980) Behaviour genetics and animal production. In: Wodzicka-Tomaszewska, M., Edey, T.N. and Lynch, J.J. (eds), *Behaviour in Relation to Reproduction, Management and Welfare of Farm Animals.* University of New England, Armidale, pp. 29–36.

Gonyou, H.W. and Stookey, J.M. (1985) Behavior of parturient ewes in group-lambing pens with and without cubicles. *Applied Animal Behaviour Science* 14, 163–71.

Grovum, W.L. and Chapman, H.W. (1988) Factors affecting the voluntary intake of food by sheep. *British Journal of Nutrition* 59, 63–72.

Grubb, P. (1974a) Social organisation of Soay sheep and the behaviour of ewes and lambs. In: Jewell, P.A., Milner, C. and Morton-Boyd, J. (eds), *Island Survivors: the Ecology of the Soay Sheep of St Kilda.* The Athlone Press, London, pp. 131–59.

Grubb, P. (1974b) Mating activity and the social significance of rams in a feral sheep community. In: Geist, V. and Walther, F. (eds), *The Behaviour of Ungulates and its Relation to Management.* IUCN, Morges, Switzerland, pp. 457–76.

Grubb, P. and Jewell, P.A. (1966) Social grouping and home range in feral Soay sheep. *Symposium of the Zoological Society of London* 18, 179–210.

Gubernick, D.J. (1981) Parent and infant attachment in mammals. In: Gubernick, D.J. and Klopfer, P.H. (eds), *Parental Care in Mammals.* Plenum Press, New York, pp. 243–305.

Hale, E.P. (1962) Domestication and the evolution of behaviour. In: Hafez, E.S.E. (ed.), *Behaviour of Domestic Animals.* Ballière Tindall & Cox, London, pp. 21–53.

Hamilton, D.R. (1974) Immunosuppressive effects of predator-induced stress on mice with acquired immunity to *Hymenolepis nana. Journal of Psychosomatic Research* 18, 145–53.

Hargreaves, A.L., and Hutson, G.D. (1990) The stress response of sheep during routine handling procedures. *Applied Animal Behaviour Science* 26, 83–90.

Harré, R. and Lamb, R. (1986) *The Dictionary of Ethology and Animal Learning*. Basil Blackwell, Oxford.

Harrison, R. (1964) *Animal Machines*. Vincent Stuart, London.

Hinch, G.N. (1989) The sucking behaviour of triplet, twin and single lambs at pasture. *Applied Animal Behaviour Science* 22, 39–48.

Hinch, G.N., Lecrivain, E., Lynch, J.J. and Elwin, R.L. (1987) Changes in maternal–young associations with increasing age of lambs. *Applied Animal Behaviour Science* 17, 305–18.

Hinch, G.N., Lynch, J.J., Elwin, R.L. and Green, G.C. (1990) Long-term associations between Merino ewes and their offspring. *Applied Animal Behaviour Science* 27, 93–103.

Hitchcock, D.K. and Hutson, G.D. (1979) The movement of sheep on inclines. *Australian Journal of Experimental Agriculture and Animal Husbandry* 19, 176–82.

Hodgson, J. (1982) Influence of sward characteristics on diet selection and herbage intake by the grazing animal. In: Hacker, J.B. (ed.), *Nutritional Limits at Animal Production from Pastures*. Commonwealth Agricultural Bureau, Farnham Royal, UK, pp. 153–66.

Hulet, C.V., Ercanrack, S.K., Price, D.A., Blackwell, R.L. and Wilson, L.O. (1962) Mating behaviour of the ram in the multi-sire pen. *Journal of Animal Science* 21, 870–4.

Hulet, C.V., Anderson, D.M., Smith, J.N. and Shupe, W.L. (1987) Bonding of sheep to cattle as an effective technique for predation control. *Applied Animal Behaviour Science* 19, 19–25.

Hume, C.W. (1962) How to befriend laboratory animals. In: *Man and Beast* (1986). Universities Federation for Animal Welfare, South Mimms, Herts, UK, pp. 6–9.

Hunter, R.F. (1964) Home range behaviour in hill sheep. In: Crisp, D.J. (ed.), *Grazing in Terrestrial and Marine Environments*. Blackwell Scientific, London, pp. 155–71.

Hunter, R.F. and Davies, G.E. (1963) The effect of method of rearing on the social behaviour of Scotish Blackface hoggets. *Animal Production* 5, 183–94.

Hurnik, J.F., Webster, A.B. and Siegel, P.B. (1985) *A Dictionary of Animal Behaviour*. University of Guelph, Guelph, Ontario, Canada.

Hutson, G.D. (1980) The effect of previous experience on sheep movement through yards. *Applied Animal Ethology* 6, 233–40.

Hutson, G.D. (1981) Sheep movement on slatted floors. *Australian Journal of Experimental Agriculture and Animal Husbandry* 21, 474–9.

Hutson, G.D. (1982) Flight distance in Merino sheep. *Animal Production* 35, 231–5.

Hutson, G.D. (1984) Spacing behaviour of sheep in pens. *Applied Animal Behaviour Science* 12, 111–19.

Hutson, G.D. (1985) The influence of barley food rewards on sheep movement through a handling system. *Applied Animal Behaviour Science* 14, 263–73.

Isaac, E. (1970) *The Geography of Domestication*. Prentice-Hall, New Jersey.

Ivins, J.D. (1952) The relative palatability of herbage plants. *Journal of British Grasslands Society* 18, 79–89.

Janeway, C.A., Bottomly, K., Horowitz, J., Kaye, J., Jones, B. and Tite, J. (1985) Modes of cell:cell interaction in the immune system. *Journal of Immunology* 135, 739s–42s.

Jankovic, B.D. (1989) Neuroimmunomodulation: facts and dilemmas. *Immunology Letters* 21, 101–18.

Jarman, P.J. (1974) The social organisation of antelope in relation to their ecology. *Behaviour* 48, 215–67.

Jewell, P.A. (1966) The concept of home range in small mammals. *Symposium of the Zoological Society of London* 18, 85–109.

Jewell, P.A. and Grubb, P. (1966) The breeding cycle, the onset of oestrus and conception in the Soay sheep. *Symposium of Zoological Society of London* 18, 224–41.

Johnson, H.M. and Torres, B.A. (1988) Immunoregulatory properties of neuroendocrine peptide hormones. *Progress in Allergy* 43, 37–67.

Jung, H.G. and Koong, L.J. (1985) Effects of hunger satiation on diet quality by grazing sheep. *Journal of Range Management* 38, 302–5.

Juwarini, E.B.H., Siebert, B.D., Lynch, J.J. and Elwin, R.L. (1981) Variation in the wheat intake of individual sheep measured by use of labelled grain: behavioural influences. *Australian Journal of Experimental Agriculture and Animal Husbandry* 21, 395–9.

Kaupfermann, I. (1985) Hypothalamus and limbic system II: motivation. In: Kandel, E.R. and Schwartz, J.H. (eds), *Principles of Neural Science*, 2nd edn Elsevier, Amsterdam, pp. 451–60.

Keller, S.E., Weiss, J.M., Schleifer, S.J., Miller, N.E. and Stein, M. (1983) Stress-induced suppression of immunity in adrenalectomised rats. *Science* 221, 1301–4.

Kelley, K.W. (1980) Stress and immune function: a bibliographic review. *Annales Recherche Vétérinaire* 11, 445–78.

Kelly, R.J., Allison, A.J. and Shackell, G.H. (1975) Libido testing and subsequent mating performance in rams. *Proceedings of the New Zealand Society of Animal Production* 35, 204–11.

Kendrick, K.M. and Baldwin, B.A. (1987) Cells in temporal cortex of conscious sheep can respond preferentially to the sight of faces. *Science* 236, 448–50.

Kenney, P.A. and Black, J.L. (1984) Factors affecting diet selection by sheep. I. Potential intake rate and acceptability of feed. *Australian Journal of Agricultural Research* 35, 551–63.

Keverne, E.B. and Kendrick, K.M. (1990) Neurochemical changes accompanying parturition and their significance for maternal behaviour. In: Krasnegor, N.A. and Bridges, R.S. (eds), *Mammalian Parenting*. Oxford University Press, Oxford, pp. 281–304.

Keverne, E.B., Levy, F., Poindron, P. and Lindsay, D.R. (1983) Vaginal stimulation: an important determinant of maternal bonding in sheep. *Science* 219, 81–3.

Key, C. and MacIver, R.M. (1980) The effects of maternal influences on sheep: breed differences in grazing, resting and courtship behaviour. *Applied Animal Ethology* 6, 33–48.

Kiley-Worthington, M. (1977) *Behavioural Problems of Farm Animals*. Oriel Press, London, p. 134.

Kilgour, R. (1972) Animal behaviour in intensive systems and its relationship to disease and production. *Australian Veterinary Journal* 48, 94–8.

Kilgour, R. (1976) Sheep behaviour: its importance in farming systems, handling, transport and pre-slaughter treatment. *Proceedings of Sheep Assembly and Transport Workshop*, W.A. Department of Agriculture, pp. 64–84.

Kilgour, R. (1985b) The behavioural background to reproduction. In: Fraser, A.F. (ed.), *Ethology of Farm Animals*. Elsevier, Oxford, pp. 279–88.

Kilgour, R. and Edey, T.N. (1977) Mating ewe lambs. *New Zealand Journal of Agriculture* 134, 10–12.

Kilgour, R. and Winfield, C.G. (1974) Sheep mating behaviour. *New Zealand Journal of Agriculture* 128, 17–22.

Kilgour, R., Pearson, A.J. and de Langen, H. (1975) Sheep dispersal patterns on hill country: techniques for study and analysis. *Proceedings of the New Zealand Society for Animal Production* 35, 191–7.

Kilgour, R.J. (1985a) Mating behaviour of rams in pens. *Australian Journal of Experimental Agriculture* 25, 298–305.

Kilgour, R.J., Barwick, S.A. and Fowler, D.G. (1985) Ram mating performance in Border Leicesters and related breed types. 2. Comparison of the performance of rams that were sexually active and inactive in pen tests. *Australian Journal of Experimental Agriculture* 25, 17–20.

Knight, T. (1983) Ram induced stimulation of ovarian and oestrous activity in anoestrous ewes–a review. *Proceedings of the New Zealand Society of Animal Production* 43, 7–11.

Krehbiel, D., Poindron, P., Levy, F. and Prud'homme, M.J. (1987) Peridural anesthesia disturbs maternal behavior in primiparous and multiparous parturient ewes. *Physiology and Behavior* 40, 463–72.

Kyriazakis, I., Emmans, G.C. and Whittemore, C.T. (1990) Diet selection in pigs: choices made by growing pigs given foods of different protein concentrations. *Animal Production* 51, 189–99.

L'Huillier, P.J., Poppi, D.P. and Fraser, T.J. (1984) Influence of green leaf distribution on diet selection by sheep and the implications for animal performance. *Proceedings of the New Zealand Society of Animal Production* 44, 105–7.

Lange, R.T. (1969) The piosphere: sheep track and dung patterns. *Journal of Range Management* 22, 396–400.

Lange, R.T. (1985) Spatial distributions of stocking intensity produced by sheep-flocks grazing Australian Chenopod shrublands. *Transactions of the Royal Society of South Australia* 109, 167–74.

Langlands, J.P. (1969) Studies of the nutritive value of the diet selected by grazing sheep. *Animal Production* 11, 369–75.

Lawrence, A.B. (1990) Mother–daughter and peer relationships of Scottish hill sheep. *Animal Behaviour* 39, 481–6.

Laycock, W.A., Buchanan, H. and Krueger, W.C. (1972) Three methods of determining diet, utilization, and trampling damage on sheep ranges. *Journal of Range Management* 25, 352–6.

Leigh, J.H. and Mulham, W.E. (1966) Selection of diet by sheep grazing semi-arid pastures on the Riverine Plain. I. A bladder saltbush (*Atriplex vesicaria*)–cotton bush (*Kochia aphylla*) community. *Australian Journal of Experimental Agriculture and Animal Husbandry* 6, 460–7.

Lenarz, M.S. (1979) Social structures and reproductive strategy in desert bighorn sheep (*Ovis canadensis mexicana*). *Journal of Mammology* 60, 671–7.

Leuthold, W. (1977) *African Ungulates: a Comparative Review of Their Ethology and Behavioural Ecology.* Springer Verlag, Berlin.

Levot, G.W., Hughes, P.B. and Kaldor, C.J. (1989) Evaluation of dressings to aid healing of mulesing wounds on sheep. *Australian Veterinary Journal* 66, 358–61.

Levy, F. and Poindron, P. (1987) The importance of amniotic fluids for the establishment of maternal behaviour in experienced and inexperienced ewes. *Animal Behaviour* 35, 1188–92.

Levy, F., Gervais, R., Kindermann, U., Litterio, M., Poindron, P. and Porter, R. (1991) Effects of early post-partum separation on maintenance of maternal responsiveness and selectivity in parturient ewes. *Applied Animal Behaviour Science* 31, 101–10.

Lincoln, G.A. and Davidson, W. (1977) The relationship between sexual and aggressive behaviour, pituitary and testicular activity during the seasonal sexual cycle of rams, and the influence of photoperiod. *Journal of Reproduction and Fertility* 49, 267–76.

Lindsay, D.R. (1966) Modification of behavioural oestrous in the ewe by social and hormonal factors. *Animal Behaviour* 14, 73–83.

Lindsay, D.R. (1985) Reproductive anomalies. In: Fraser, A.F. (ed.), *Ethology of Farm Animals.* Elsevier, Oxford, pp. 423–9.

Lindsay, D.R. (1988) Reproductive behaviour in survival: a comparison between wild and domestic sheep. *Australian Journal of Biological Science* 41, 97–102.

Lindsay, D.R. and Ellsmore, J. (1968) The effect of breed, season and competition on mating behaviour of rams. *Australian Journal of Experimental Agriculture and Animal Husbandry* 8, 649–52.

Lindsay, D.R., Dunsmore, D.G., Williams, J.D. and Syme, G.J. (1976) Audience effects on mating behaviour of rams. *Animal Behaviour* 24, 818–21.

Livingston, P.B. (1985) Neurophysiology. In: West, J.B. (ed.), *Best and Taylor's Physiological Basis of Medical Practice,* 11th edn. Williams & Wilkins, Baltimore, pp. 970–1297.

Lobato, J.F.P., Pearce, G.R. and Beilharz, R.G. (1980) Effect of early familiarisation with dietary supplements on the subsequent ingestion of mollasses-urea blocks by sheep. *Applied Animal Ethology* 6, 149–61.

Lyford, S.J. (1988) Growth and development of the ruminant digestive system. In: Church, D.C. (ed.), *The Ruminant Animal.* Prentice-Hall, New Jersey, pp. 44–63.

Lynch, J.J. (1974) Merino sheep: some factors affecting their distribution in very large paddocks. In: Geist, V. and Walther, F. (eds), *The Behaviour of Ungulates in Relation to Management,* IUCN, Morges, Switzerland, pp. 697–707.

Lynch, J.J. (1977) Movement of some rangeland herbivores in relation to their feed and water supply. In: *The Impact of Herbivores on Arid and Semi-arid Rangelands.* Proceedings of the 2nd United States/Australia Rangeland Panel, Adelaide, 1972, Australian Rangeland Society, Perth, Western Australia, pp. 163–72.

Lynch, J.J. and Alexander, G. (1977) Sheltering behaviour of lambing Merino sheep in relation to grass hedges and artificial windbreaks. *Australian Journal of Agricultural Research* 28, 691–701.

Lynch, J.J., Woodgush, D.G.M. and Davies, H.I. (1985) Aggression and nearest neighbours in a flock of Scottish Blackface sheep. *Biology of Behaviour* 10, 215–25.

Lynch, J.J. Hinch, G.N., Bouissou, M.F., Elwin, R.L., Green, G.C. and Davies, H.I. (1989) Social organisation in young Merino and Merino × Border Leicester ewes. *Applied Animal Behaviour Science* 22, 49–63.

Lynch, J.J., Leng, R.A., Hinch, G.N., Nolan, J., Bindon, B.M. and Piper, L.R. (1990) Effects of cottonseed supplementation on birthweight and survival of lambs from a range of litter sizes. *Proceedings of Australian Society of Animal Production* 18, 516.

MacArthur, R.A., Johnston, R.H. and Geist, V. (1979) Factors influencing the heart rate in free ranging bighorn sheep: a physiological approach to the study of wildlife harassment. *Canadian Journal of Zoology* 57, 2010–21.

MacLean, P.D. (1970) The limbic brain in relation to the psychoses. In: Black, P. (ed.), *Physiological Correlates of Emotion.* Academic Press, New York, pp. 130–46.

Manning, A. (1979) *An Introduction to Animal Behaviour*, 3rd edn, Arnold, London.

Mason, I.L. (1969) *A World Dictionary of Livestock Breeds, Types and Varieties*, 2nd edn, Commonwealth Agricultural Bureaux, Farnham Royal, UK.

Mattner, P.E., Braden, A.W.H. and Turnbull, K.E. (1967) Studies of flock mating of sheep. 1. Mating behaviour. *Australian Journal of Experimental Agriculture and Animal Husbandry* 7, 102–9.

Mattner, P.E., Braden, A.W.H. and George, J.M. (1971) Studies in flock mating of sheep. 4. The relation of libido tests to subsequent service activity of young rams. *Australian Journal of Experimental Agriculture and Animal Husbandry* 11, 473–7.

McBride, G. (1971) Theories of animal spacing: the role of flight, fight and social distance. In: Esser, A.H. (ed.), *Behaviour and Environment – The Use of Space by Animals and Man.* Plenum Press, New York and London, pp. 53–68.

McClelland, B.E. (1991) Courtship and agonistic behaviour in mouflon sheep. *Applied Animal Behaviour Science* 29, 67–85.

McFarland, D. (1985) *Oxford Companion to Animal Behaviour.* Oxford University Press, Oxford.

McFarland, D.J. (1986) Frustration. In: Harré, R. and Lamb, R. (eds), *The Dictionary of Ethology and Animal Learning.* Basil Blackwell, Oxford.

Mellor, D.J. and Murray, L. (1989) Effects of tail-docking and castration on behaviour and plasma cortisol concentrations in young lambs. *Research in Veterinary Science* 46, 387–91.

Meyer, H.H. (1979) Ewe and teaser breed effects on reproductive behaviour and performance. *Proceedings of the New Zealand Society of Animal Production* 39, 68–76.

Mirza, S.N. and Provenza, F.D. (1992) Effects of age and conditions of exposure on maternally mediated food selection in lambs. *Applied Animal Behaviour Science* (in press).

Morton, D.B. and Griffiths, P.H.M. (1985) Guidelines on the recognition of pain, distress and discomfort in experimental animals and an hypothesis for assessment. *Veterinary Record* 116, 431–6.

Mottershead, B.E., Lynch, J.J., Elwin, R.L. and Green, G.C. (1985) A note on the acceptance of several types of cereal grain by young sheep with and without prior experience of wheat. *Animal Production* 41, 257–9.

Naaktgeboren, C. (1979) Behavioural aspects of parturition. *Animal Reproduction Science* 2, 155–66.

Naess, A. (1989) *Ecology, Community and Lifestyle: Outline of an Ecophilosophy.* Cambridge University Press, Cambridge.

Nichols, J.E. (1944) The behaviour of sheep grazing during drought in Western Australia. In: *Animal Behaviour.* Joint Meeting with the Institute for the Study of Animal Behaviour, 25 October 1949, British Society for Animal Production, London, pp. 66–73.

Nowak, R. (1988) Relationship between neonatal mortality and mother recognition by 12-hour-old Merino lambs. *Proceedings of the Australian Society of Animal Production* 17, 449.

Nowak, R. (1990) Lamb's bleats: important for the establishment of the mother-young bond? *Behaviour* 115, 14–29.

Nowak, R. and Lindsay, D.R. (1990) Effect of genotype and litter size on discrimination of mothers by their twelve-hour-old lambs. *Behaviour* 115, 1–13.

Obst, J.M. and Ellis, J.V. (1977) Weather, ewe behaviour and lamb mortality. *Agricultural Record* 4, 44–9.

Oppenheim, J.J. and Shevach, E.M. (eds) (1990) *Immunophysiology.* Oxford University Press, New York.

Owens, J.L., Bindon, B.M., Edey, T.N. and Piper, L.R. (1980) Neonatal behaviour in high fecundity Booroola Merino ewes. In: Wodzicka-Tomaszewska, M., Edey, T.N. and Lynch, J.J. (eds), *Reviews in Rural Science No. 4. Behaviour in Relation to Reproduction, Management and Welfare.* University of New England Publishing Unit, Armidale, Australia, pp. 113–16.

Owens, J.L., Bindon, B.M., Edey, T.N. and Piper, L.R. (1985) Behaviour at parturition and lamb survival of Booroola Merino sheep. *Livestock Production Science* 13, 359–72.

Paterson, I.W. and Coleman, C.D. (1982) Activity patterns of seaweed-eating sheep on North Ronaldsay, Orkney. *Applied Animal Ethology* 8, 137–46.

Payan, D.G., McGillis, J.P. and Goetzl, E.J. (1986) Neuroimmunology. *Advances in Immunology* 39, 299–323.

Penning, P.D., Parsons, A.J., Orr, R.J. and Treacher, T.T. (1991) Intake and behaviour responses by sheep to changes in sward characteristics under continuous stocking. *Grass and Forage Science* 46, 15–28.

Pfister, J.A., Müller-Schwarze, D. and Balph, D.F. (1990) Effects of predator fecal odors on feed selection by sheep and cattle. *Journal of Chemical Ecology* 16(2), 323–9.

Plaut, S.M. and Friedman, S.B. (1982) Stress, coping behavior and resistance to disease. *Psychotherapeutica and Psychosomatica* 38, 274–83.

Plutchick, R. (1980) *Emotion: a Psychoevolutionary Synthesis.* Harper & Row, New York, pp. 41–3.

Plutchick, R. (1986) Emotion. In: Harré, R. and Lamb, R. (eds), *The Dictionary of Ethology and Animal Learning.* Basil Blackwell, Oxford.

Poindron, P. and Le Neindre, P. (1980) Endocrine and sensory regulation of maternal behavior in the ewe. In: *Advances in the Study of Behaviour.* Academic Press, pp. 75–99.

Poindron, P. and Levy, F. (1990) Physiological, sensory and experiential determinants of maternal behaviour in sheep. In: Krasnegor, N.A. & Bridges, R.S.

(eds), *Mammalian Parenting*. Oxford University Press, Oxford, pp. 133–56.

Poindron, P., Le Neindre, P., Raksanyi, I., Trillat, G. and Orgeur, P. (1980) Importance of the characterstics of the young in the manifestation and establishment of maternal behaviour in sheep. *Reproduction Nutrition and Development* 20, 817–26.

Poindron, P., Levy, F. and Krehbiel, D. (1988) Genital, olfactory and endocrine interactions in the development of maternal behaviour in the parturient ewe. *Psychoneuroendocrinology* 13, 99–125.

Price, E.O. (1984) Behavioural aspects of animal domestication. *Quarterly Review of Biology* 59, 1–32.

Price, E.O., Estep, D.Q., Wallach, S.J.R. and Dally, M.R. (1991a) Sexual performance of rams as determined by maturation and sexual experience. *Journal of Animal Science* 69, 1047–52.

Price, E.O., Wallach, S.J.R. and Dally, M.R. (1991b) Effects of sexual stimulation on the sexual performance of rams. *Applied Animal Behaviour Science* 30, 333–40.

Provenza, F.D. and Balph, D.F. (1990) Applicability of five diet-selection models to various foraging challenges ruminants encounter. In: Hughes, R.N. (ed.), *Behavioural Mechanisms of Food Selection*. NATO ASI Series G: Ecological Sciences, Vol. 20. Springer Verlag, Heidelberg and New York, pp. 423–59.

Provenza, F.D., Pfister, J.A. and Cheney, C.D. (1992) Mechanisms of learning in diet selection with reference to phytotoxicosis in herbivores. *Journal of Range Management* 45, 36–45.

Putu, I.G., Poindron, P. and Lindsay, D.R. (1988) Early disturbance of Merino ewes from the birth site increases lamb separations and mortality. *Proceedings of the Australian Society of Animal Production* 17, 298–301.

Raadsma, H.W. and Edey, T.N. (1985) Mating performance of paddock-mated rams. II. Changes in sexual and general activity during the joining period. *Animal Reproduction Science* 8, 101–7.

Riley, V.E. (1981) Psychoneuroendocrine influences on immunocompetence and neoplasia. *Science* 212, 1100–9.

Robinson, T.J. (1955) Quantitative studies on hormonal induction of oestrus in spayed ewes. *Journal of Endocrinology* 12, 166–73.

Russell, W.M.S. and Burch, R.L. (1959) *The Principles of Humane Experimental Technique*. Methuen, London.

Ryder, M.L. (1983) *Sheep and Man*. Duckworth, London.

Ryder, M.L. (1984) Sheep. In: Mason, I.L. (ed.), *Evolution of Domesticated Animals*. Longman, London, pp. 63–85.

Ryle, G. (1966) *Plato's Progress*. Cambridge University Press, Cambridge.

Sanford, J., Ewbank, R., Molony, V., Tavernor, W.D. and Uvarov, O. (1986) Guidelines for the recognition and assessment of pain in animals: prepared by a working party of the Association of Veterinary Teachers and Research Workers. *Veterinary Record* 118, 334–8.

Schaller, G.B. (1977) *The Mountain Monarchs: Wild Sheep and Goats in the Himalayas*. University of Chicago Press, Chicago.

Schaller, G.B. and Miraz, Z.B. (1974) On the behaviour of Punjab Urial (*Ovis orientalis punjabiensis*). In: Geist, V. and Walther, F. (eds), *The Behaviour of Ungulates in Relation to Management*. IUCN Morges, Switzerland, pp. 306–23.

Schinckel, P.G. (1954) The effect of the ram on the incidence and occurrence of oestrus in ewes. *Australian Veterinary Journal* 30, 189–95.

Selye, H. (1946) The general adaptation syndrome and the diseases of adaptation. *Journal of Clinical Endocrinology* 6, 117–230.

Selye, H. (1976) *Stress in Health and Disease.* Butterworth, London.

Senate Select Committee on Animal Welfare (1990) *Sheep.* Australian Government Publishing Service, Canberra, Australia.

Shackleton, D.M. (1991) Social maturation and productivity in bighorn sheep: are young males incompetent? *Applied Animal Behaviour Science* 29, 173–84.

Shackleton, D.M. and Shank, C.G. (1984) A review of the social behaviour of feral and wild sheep and goats. *Journal of Animal Science* 58, 500–9.

Sherwin, C.M. and Johnson, K.G. (1987) The influence of social factors on the use of shade by sheep. *Applied Animal Behaviour Science* 18, 143–55.

Shillito, E.E. (1975) A comparison of the role of vision and hearing in lambs finding their own dams. *Applied Animal Ethology* 1, 369–77.

Shillito, E. and Alexander, G. (1975) Mutual recognition amongst ewes and lambs of four breeds of sheep (*Ovis aries*). *Applied Animal Ethology* 1, 151–65.

Shillito-Walser, E.E. (1978) A comparison of the role of vision and hearing in ewes finding their own lambs. *Applied Animal Ethology* 4, 71–9.

Shillito-Walser, E. and Alexander, G. (1980) Mutual recognition between ewes and lambs. *Reproduction, Nutrition, Development* 20, 807–16.

Shillito-Walser, E.E. and Hague, P. (1980) Variations in the structures of bleats from sheep of four different breeds. *Behaviour* 75, 22–35.

Shillito-Walser, E.E., Hague, P. and Yeomans, M. (1983) Variations in the strength of maternal behaviour and its conflict with flocking behaviour in Dalesbred, Jacob and Soay ewes. *Applied Animal Ethology* 10, 245–50.

Shillito-Walser, E.E., Walters, E. and Ellison, J. (1982) Observations on vocalisation of ewes and lambs in the field. *Behaviour* 91, 190–203.

Shillito-Walser, E., Willadson, S. and Hague, P. (1981) Pair association between lambs of different breeds born to Jacob and Dalesbred ewes after embryo transplantation. *Applied Animal Ethology* 1, 351–8.

Shillito-Walser, E.E., Walters, E., Hague, P. and Williams, T. (1985) Responses of lambs to model ewes. *Behaviour* 95, 110–20.

Shutt, D.A., Fell, L.R., Connell, R., Bell, A.K., Wallace, C.A. and Smith, A.I. (1987) Stress-induced changes in plasma concentrations of immunoreactive β-endorphin and cortisol in response to routine surgical procedures in lambs. *Australian Journal for Biological Science* 40, 97–103.

Shutt, D.A., Fell, L.R., Connell, R. and Bell, A.K. (1988) Stress responses in lambs docked and castrated surgically or by the use of rubber rings. *Australian Veterinary Journal* 65, 5–7.

Signoret, J.P. (1980) Endocrine basis of reproductive behaviour in female domestic animals. In: Wodzicka-Tomaszewska, M., Edey, A.N. and Lynch, J.J. (eds), *Behaviour in Relation to Reproduction, Management and Welfare of Farm Animals*, University of New England, Armidale, pp. 3–9.

Singer, P. (1976) *Animal Liberation: A New Ethic for Our Treatment of Animals.* Jonathan Cape, London.

Slee, J. and Springbett, A. (1986) Early post-natal behaviour in lambs of ten breeds. *Applied Animal Behaviour Science* 15, 229–40.

Slee, J., Griffiths, R.G. and Samson, D.E. (1980) Hypothermia in newborn lambs induced by experimental immersion in a water bath and by natural exposure outdoors. *Research in Veterinary Science* 28, 275–80.

Squires, V. (1981) *Livestock Management in the Arid Zone.* Inkata Press, Melbourne.

Squires, V.R. and Daws, G.T. (1975) Leadership and dominance in Merino and Border-Leicester sheep. *Applied Animal Ethology* 1, 263–74.

Stevens, D., Alexander, G. and Lynch, J.J. (1981) Do Merino ewes seek isolation or shelter at lambing? *Applied Animal Ethology* 7, 149–55.

Stevens, D., Alexander, G. and Lynch, J.J. (1982) Lamb mortality due to inadequate care of twins by Merino ewes. *Applied Animal Ethology* 8, 243–52.

Stolba, A. and Wood-Gush, D.G.M. (1984) The identification of behavioural key features and their incorporation into a housing design for pigs. *Annales Recherche Vétérinaire* 15, 287–98.

Stolba, A., Hinch, G.N., Lynch, J.J., Adams, D.B., Munro, R.K. and Davies, H.I. (1990) Social organisation of Merino sheep of different ages, sex and family structure. *Applied Animal Behaviour Science* 27, 337–49.

Syme, L.A. (1981) Social disruption and forced movement orders in sheep. *Animal Behaviour* 29, 283–8.

Syme, G.J. and Syme, L.A. (1979) *Social Structure in Farm Animals.* Elsevier, Amsterdam.

Synnott, A.L. and Fulkerson, W.J. (1984) Influence of social interaction between rams on their serving capacity. *Applied Animal Behaviour Science* 11, 283–9.

Taylor, J.A., Robinson, G.G., Hedges, D.A. and Whalley, R.D.B. (1987) Camping and faeces distribution by Merino sheep. *Applied Animal Behaviour Science* 17, 273–88.

Tepper, B.J. and Kanarek, R.B. (1989) Selection of protein and fat by diabetic rats following separate dilution of the dietary sources. *Physiology and Behaviour* 45, 49–61.

Thorhallsdottir, A.G., Provenza, F.D. and Balph, D.F. (1987) Food aversion learning in lambs with or without a mother, discrimination, novelty and persistence. *Applied Animal Behaviour Science* 18, 327–40.

Thorhallsdottir, A.G., Provenza, F.D. and Balph, D.F. (1990) Ability of lambs to learn about novel foods while observing or participating with social models. *Applied Animal Behaviour Science* 25, 25–33.

Thorpe, W.H. (1962) *Biology and the Nature of Man.* Oxford University Press, Oxford.

Thwaites, C.J. (1982) Development of mating behaviour in the prepubertal ram. *Animal Behaviour* 30, 1053–9.

Tilbrook, A.J. (1987a) Physical and behavioural factors affecting the sexual attractiveness of the ewe. *Applied Animal Behaviour Science* 17, 109–15.

Tilbrook, A.J. (1987b) The influence of factors associated with oestrus on the sexual attractiveness of ewes to rams. *Applied Animal Behaviour Science* 17, 117–28.

Tilbrook, A.J. and Cameron, A.W.N. (1989) Ram mating preferences for woolly rather than shorn ewes. *Applied Animal Behaviour Science* 24, 301–12.

Tomkins, T. and Bryant, M.J. (1974) Oestrous behaviour of the ewe and the influence of treatment with progestagen. *Journal of Reproduction and Fertility* 41, 121–32.

Tribe, D.E. (1950) Social facilitation in grazing sheep. *Nature* 166, 74.

UFAW (1987) *Humane Slaughter of Animals for Food.* Universities Federation for Animal Welfare, South Mimms, Herts.

Van Rooijen, J. (1990) Physiology, behaviour and wellbeing. *Applied Animal Behaviour Science* 27, 367–8.

Vince, M.A. and Billing, A.E. (1986) Infancy in the sheep: the part played by sensory stimulation in bonding between the ewe and lamb. In: Lipsitt, L.P. and Rovee-Collier, C. (eds), *Advances in Infancy Research*, Vol. IV. Ablex Norwood, New Jersey, pp. 1–37.

Vince, M.A., Armitage, S.E., Baldwin, B.A. and Toner, J. (1982) The sound environment of the foetal sheep. *Behaviour* 91, 296–315.

Vince, M.A., Lynch, J.J., Mottershead, B., Green, G. and Elwin, R. (1985) Sensory factors involved in immediately postnatal ewe/lamb bonding. *Behaviour* 94, 60–84.

Vince, M.A., Lynch, J.J., Mottershead, B.E., Green, G.C. and Elwin, R.L. (1987) Interactions between normal ewes and newly born lambs deprived of visual, olfactory and tactile sensory information. *Applied Animal Behaviour Science* 19, 119–36.

Vines, G. (1981) Wolves in dog's clothing. *New Scientist* 91, 648–52.

Wall, P.D. (1979) On the relation of injury to pain. *Pain* 6, 253–64.

Warren, J.T. and Mysterud, I. (1991) Summer habitat use and activity patterns of domestic sheep on coniferous forest range in southern Norway. *Journal of Range Management* 44, 2–6.

Weinmann, C.J. and Rothman, A.H. (1967) Effect of stress on acquired immunity to the dwarf tapeworm *Hymenolepis nana. Experimental Parasitology* 21, 61–7.

Weston, R.H. and Poppi, D.P. (1987) In: Hacker, J.B. and Ternouth, J.H. (eds), *Nutrition of Herbivores.* Academic Press, Sydney, pp. 133–61.

Whateley, J., Kilgour, R. and Dalton, D.C. (1974) Behaviour of hill country sheep breeds during farming routines. *Proceedings of the New Zealand Society of Animal Production* 34, 28–36.

Wilson, P.R. and Orwin, D.F.C. (1964) The sheep population of Campbell Island. *New Zealand Journal of Science* 7, 460–90.

Winfield, C.G. and Kilgour, R. (1977) The mating behaviour of rams in a pedigree pen-mating system in relation to breed and fertility. *Animal Production* 24, 197–201.

Winfield, C.G. and Mullaney, P.D. (1973) A note on the social behaviour of a flock of Merino and Wiltshire horn sheep. *Animal Production* 17, 93–5.

Wodzicka-Tomaszewska, M., Kilgour, R. and Ryan, M. (1981) 'Libido' in the larger farm animal: a review. *Applied Animal Behaviour Science* 7, 203–22.

Woolliams, C., Wiener, G. and Macleod, N.S.M. (1983) The effects of breed, breeding system and other factors on lamb mortality. 3. Factors influencing the incidence of weakly lambs as a cause of death. *Journal of Agricultural Science, Cambridge* 100, 563–70.

Zenchak, J.J. and Anderson, G.C. (1980) Sexual performance levels of rams (*Ovis aries*) as affected by social experiences during rearing. *Journal of Animal Science* 50, 167–74.

Index of Sheep Breeds

Index